"Doug Kenrick is an ideal person to bring the real evolutionary psychology to the attention of a general audience. Not only has he done some outstanding research in this area but he is superb writer with that rarest of academic attributes—a sense of humor that translates well to the page. Readers of his book will enjoy themselves, unless they are grumpy academicians who have unfairly and inaccurately concluded that they do not have to pay attention to evolutionary psychology. The rest of us will have a fine time as we learn just how much evolutionary psychologists have to contribute to an understanding of our species."

—John Alcock, Regents Professor of Life Sciences, Arizona State University; author of *The Triumph of Sociobiology*, *Sonoran Desert Summer*, and *Animal Behavior*

"When one thinks about evolutionary psychology—which is increasingly necessary in all of the behavioral sciences—one has to think of Douglas Kenrick. Because of his rarified abilities to identify, uncover, and communicate important answers to vital human questions, he is both at the center and at the top of this burgeoning scientific field."

—Robert Cialdini, Regents Professor of Psychology and Carey Distinguished Professor of Business, Arizona State University; author of *New York Times* best-selling *Influence*

"Doug Kenrick is a brilliant thinker, a brilliant researcher, and a brilliant writer. His scientific discoveries dazzle and mesmerize, but they are also seminal. One testament is Kenrick's high professional profile. In my evolutionary psychology textbook, Kenrick's work is cited more than any other scientist. To top it off, Kenrick has a phenomenal sense of humor, unmatched by any other academic I've seen or heard; he could have easily been a professional stand-up comic. More to the point, he brings his wit and intellectual flair to dazzle readers, making the science of the human mind spring to life and grab readers around the throat. It's a certainty that Kenrick's book will be brilliant, witty, controversial, and a good bet to hit the best-seller list."

—David Buss, professor of psychology, University of Texas; author of *The Evolution of Desire* and *The Murderer Next Door*

"Kenrick is exceptional among psychologists in understanding the connections between our biological evolution and our brain processes, but he is also exceptional among evolutionary psychologists in applying this knowledge to issues that really matter—like who we love and who we hate, why people kill and save others, why they want meaning rather than chaos."

—Pascal Boyer, Henry Luce Professor of Anthropology
and Psychology, Washington University;
author of *Religion Explained*

"Doug Kenrick understood the meaning and significance of unconscious evolutionary drivers of social behavior long before that line of thinking became fashionable—his work on fundamental social motives, published in the top psychology research journals such as *Psychological Review* and *Behavioral and Brain Sciences*, has been a tremendously influential integration of evolutionary psychology, cognitive science, and the unconscious. This provocative and fascinating book will be as well."

—John Bargh, professor of psychology and cognitive science,
Yale University; editor of *The New Unconscious*

"Doug Kenrick is perfectly suited to write this kind of book: One that is intellectually dead-on accurate with respect to facts and exposition of theory, and yet engaging and thought-provoking to a wide audience. Kenrick is a leading scholar and researcher in the field of evolutionary psychology. He knows the field as well as anyone. He thinks deeply, critically, and clearly. At the same time, he's a brilliant writer and teacher—and as able to be engagingly witty as anyone I know. His wit may focus on some sensitive topics, but it's smart, intellectually insightful, and brilliantly illuminating."

—Steven Gangestad, Distinguished Professor of
Psychology, University of New Mexico; author of *The
Evolutionary Biology of Human Female Sexuality*

"His engaging and humorous writing style, and his fitting use of interesting and captivating real-life examples, makes even the most complex scientific findings come alive. Thus, his book is both scholarly and entertaining, a relatively rare combination in today's marketplace."

—Jeffry Simpson, editor of *Journal of Personality & Social
Psychology*; professor of psychology, University of
Minnesota; author of *The Evolution of Mind*

SEX, MURDER, AND THE MEANING OF LIFE

Also by Douglas T. Kenrick

Social Psychology (with Steven L. Neuberg and Robert B. Cialdini)

Evolutionary Social Psychology (with Jeffry A. Simpson)

SEX,
MURDER,
AND THE
MEANING OF LIFE

A Psychologist Investigates How
Evolution, Cognition, and Complexity are
Revolutionizing Our View of Human Nature

DOUGLAS T. KENRICK

BASIC BOOKS
A Member of the Perseus Books Group
New York

Published by Basic Books,
A Member of the Perseus Books Group

Books published by Basic Books are available at special discounts for bulk purchases in the United States by corporations, institutions, and other organizations. For more information, please contact the Special Markets Department at the Perseus Books Group, 2300 Chestnut Street, Suite 200, Philadelphia, PA 19103, or call (800) 810-4145, ext. 5000, or e-mail special.markets@perseusbooks.com.

Designed by Brent Wilcox

Library of Congress Cataloging-in-Publication Data
Kenrick, Douglas T.
   Sex, murder, and the meaning of life : a psychologist investigates how evolution, cognition, and complexity are revolutionizing our view of human nature / Douglas T. Kenrick.
      p.   cm.
   Includes bibliographical references and index.
   ISBN 978-0-465-02044-7 (hardcover : alk. paper)
   ISBN 978-0-465-02342-4 (e-book)
   1. Cognitive science.   2. Evolution (Biology)   3. Human evolution.
4. Social psychology.   5. Sex.   6. Aggressiveness.   I. Title.
BF311.K386 2011
153—dc22

2010040963

10 9 8 7 6 5 4 3 2 1

Contents

Introduction

YOU, ME, CHARLES DARWIN, AND DR. SEUSS

You and I have probably never met, but you might be shocked to learn how well we know one another and how intimately our lives are connected.

You might be especially surprised at this statement if you were to make a casual comparison of our lives. For example, maybe your grandparents are Italian or Iranian or Lithuanian, with absolutely no link to my shantytown Irish ancestors. Maybe all your relatives have been law-abiding accountants or police officers, unlike my father and brother, who both served time in Sing Sing, or my uncle, who was reputedly a mobster. Maybe you never had a stepfather, or if you did, maybe you never had the slightest thought about murdering him. Maybe you had a consistent and spotless academic record; whereas I was expelled from two high schools and almost tossed out of a community college before somehow flipping around to become a university professor who writes scientific articles about human behavior (a fact that still surprises even me, and would probably shock a few of the teachers who awarded me well-deserved failing grades). Maybe you never had the surprisingly uncomfortable experience of watching yourself as a talking head on scientific documentaries or on *The Oprah Winfrey Show*, or if you had, maybe you discussed English literature in complete eloquent sentences with impeccable Oxford enunciation (unlike my, uh, New Yawk accent). And at the most trivial level, maybe the musical module in your mind neatly encompasses the complete works of Brahms, Beethoven, and Stravinsky, whereas my musical mind is an eclectic hodgepodge of Dion and the Belmonts, the Electric Prunes, Sopwith Camel, Ali Farka Touré, and Panjabi MC.

Despite our differences in family background, education, occupation, and musical exposure, though, I'll stick by my claim that we are intimately connected. We share a common human nature. However unique your upbringing or mine might be, if we were magically switched, *Prince and the Pauper*-style, we would probably respond in surprisingly similar ways to one another's situations. In this book, I'll explore revolutionary recent developments in evolutionary biology and cognitive science to clarify those connections.

You, me, Jennifer Lopez, and the old Mongolian fellow now walking down a back street in Ulaanbaatar are connected by more than just a common evolutionary past, though. Every day, your decisions and mine feed into a network of social influence that links us not only to our immediate neighbors but also to stockbrokers on Wall Street and to total strangers halfway around the world. Indeed, all human beings are interconnected in a complex web, like millions of ants in a giant colony. An emerging scientific revolution known as complexity theory neatly explains how all that works. Combined with the insights of evolutionary biology and cognitive science, as you will see, the science of complexity gives us a whole new understanding of what it means to be a member of the human race.

What This Book Is About, and the Cheat-Sheet Summary

Despite what you might have read in Malcolm Gladwell's book *Blink*, first impressions can be misleading. If you do a blink-style speed-read of this book, you might think it is mostly about me. I do in fact open each chapter with a personal experience: I started studying the downside of ogling beautiful women because I wasted a good portion of my student years doing just that. I began examining conspicuous consumption and irrational economic decision-making because I had done plenty of both. And I started doing research on homicidal fantasies after having a few of my own. But if you keep reading, I am pretty sure you'll discover that this book is really about you, your family, and your friends and about the important decisions you confront every day.

Likewise, if you read only the first few chapters, you might be misled into thinking this is just a book about the evolutionary psychology of sex, violence, and prejudice. In the later chapters, though, I will spell out how the self-centered psychology of sex, violence, and prejudice are intimately connected to the other-centered psychology of family values, religion, politics, and global economics.

Indeed, this is a book about the biggest question we can ask: What is the meaning of life? When we ask that question, we are sometimes asking how life, the universe, and everything fit together. By combining a few modern scientific insights into evolution, cognition, and complexity, we can now actually begin to answer that grand question. More often, though, what we want to know is, "How can I live a more meaningful life?" That is also a critically important question, one that leads many people to read self-help books, join religious groups, learn to meditate, or enter psychoanalysis. Academic intellectuals who think Big Thoughts about scientific integration usually avoid speculations about the "how-to" version of the meaning question. They leave that to the fuzzy-headed gurus who write pop psych books about Zen and the art of belly-button contemplation. But I think the scientific lessons we have learned about the coherence of nature may have something very important to teach us about how to live a more meaningful life.

Although this is a book about big scientific ideas, it is also about the fun side of solving intellectual mysteries—a frolicking journey to visit the wild things inside the human mind and a jolly ride back in time for dinner. It is not a college textbook, and there will not be a quiz at the end. But for those who like to read the summary before reading the text, here are the five key elements of the story that would go onto the flash cards:

1. *Simple selfish rules.* By studying human behavior in an evolutionary context, we have discovered an array of simple and selfish rules underlying our everyday decisions. The old view was that those rules only applied to sex and aggression and that evolutionary analyses did not apply to more complex decisions. But I will discuss exciting new findings that tie the same set of rules to the whole range of human behaviors, including artistic creativity, economic consumption, religion, and politics, as well as the more nuts-and-bolts aspects of courtship and sex.

2. *Simple rules do not mean simple people.* Contrary to popular opinion, the evolved decision rules inside our brains are not rigid; instead, they are flexibly tuned to the environment. Work from my lab reveals that we are all multiple personalities; that is, each one of us can shift among several different *subselves*, each capable of adaptively changing the way we think and behave, to negotiate the qualitatively different threats and opportunities that pop up in seven key domains of social life. As I describe in Chapter 6, I have dubbed those subselves the *team player* (concerned with the goal of making friends), the *go-getter* (concerned with getting ahead), the *night watchman* (concerned with protecting us from the bad guys), the *compulsive* (concerned with protecting us from disease), the *swinging single* (concerned with finding mates), the *good spouse* (concerned with the very

different problem of keeping those mates), and the *parent* (concerned with taking care of our kin, especially any children we might have). These different subselves come on line at different times of our lives, and, as I will describe in Chapter 7, thinking about their links to fundamental goals led me to rebuild Abraham Maslow's classic pyramid of human motives.

3. *Simple does not mean irrational.* Although our default decision rules sometimes lead us to behave in ways that seem irrational, other recent work from our lab indicates that the simple rules themselves manifest what I call *deep rationality.* Underneath our apparently irrational judgments, we are a lot smarter than even the most rational economists ever dreamed. I describe this new approach to economic psychology in Chapter 9 ("Peacocks, Porsches, and Pablo Picasso") and Chapter 11 ("Deep Rationality and Evolutionary Economics").

4. *Selfish rules do not create selfish people.* Although they serve selfish ends, simple decision rules do not necessarily inspire us toward self-centered behavior. Instead, the rules inside our individual heads are exquisitely calibrated to help us fit in smoothly with other people. In the book's final pages, I will describe how this new approach completely overturns people's stereotypical assumptions about the lessons of evolutionary psychology for our relationships with our friends, lovers, and family members. I will also talk about how this new view gave me a personal insight into the way to live a more meaningful life.

5. *Simple rules unfold into societal complexity.* Amazingly, all the complexities of human society—religious and political movements, economic markets, and more—emerge out of the dynamic interaction of the simple rules operating inside individual people's heads. I describe how all that works in Chapter 12 ("Bad Crowds, Chaotic Attractors, and Humans as Ants").

Procrastination 101

I first thought of writing a book of popular science more than thirty-five years ago. More than two decades passed before I started writing the volume now before you, for which I drafted a first chapter nine years ago. Partly the delay had to do with the demands of my work; it takes time to prepare lectures, apply for grants, design and conduct experiments, and publish papers on the results. But the truth is that I spent the better part of those three decades procrastinating.

In the long run, my procrastination has turned out to be a good thing. When I want to procrastinate, I don't just sit around watching reruns of old television shows; I sneak off to the bookstore, where I search for a book that has absolutely nothing to do with my current projects. Some of the books I've stumbled on were scientific ones, by brilliant researchers I've gotten to know, and sometimes work with, over those years, including John Alcock, David Buss, Steven Pinker, Geoffrey Miller, and Sonja Lyubomirsky. You will see some of their ideas as my story unfolds.

Not all my procrastination is so virtuous, though. There is a second category of books I read when my goal is *pure* procrastination: autobiographies of people I had never heard of before. Some of my favorite such distractions have come from Anthony Bourdain's *Kitchen Confidential*, Mary Karr's *The Liars' Club*, and Robert Sapolsky's *A Primate's Memoir*. From them, I have gotten glimpses into corners of the world I have never been able to visit and lessons for my own life that I could not have learned otherwise. Besides that, my attraction to that kind of up-close account led me to tell this story in a more personal way, describing the links between scientific research and puzzling events in my own life, from minor irrationalities in economic decision-making to those homicidal fantasies and high school expulsions.

Finally, because I have two sons (one born at the beginning and another toward the end of the three decades I've been working on this book), I've also read aloud most of the collected works of Dr. Seuss, Douglas Adams, and Mark Twain. In what follows, I hope you will find a satisfying fusion of these different influences—a superficially personal adventure that overlays a deeper, more universally relevant argument. And I hope you will also discover that the particulars of this story include a general lesson or two that apply to your own adventure. Be forewarned: There is sex and violence in here, so even though this book has a happy ending, I do not recommend you read this one out loud to your kids.

Chapter 1

STANDING IN THE GUTTER

In 1975, the world was about to end. The Jehovah's Witnesses had predicted Armageddon, and signs of cataclysmic change were everywhere. The North Vietnamese army drove the last American soldiers out of Saigon, Indira Gandhi suspended civil liberties in India, and the Provisional Irish Republican Army bombed the London Hilton. In the United States, members of a militant radical group who called themselves the Weathermen were bombing banks, corporation headquarters, and the State Department. A former U.S. attorney general and several leading White House officials were being hauled off to prison. There were two attempts to assassinate President Gerald Ford within seventeen days, one by a disciple of mass murderer Charles Manson. Elvis Presley, the only real king most Americans have ever recognized, was on a fast track to self-destruction. Oblivious to the coming end of the world and to the sound of falling kings and world leaders, unconcerned young people strutted lasciviously in polyester disco outfits to the sounds of KC and the Sunshine Band's "Get Down Tonight."

I was a bit out of touch with all that chaos, because I spent the best part of that momentous year nestled away in either the library or the psychology lab. But like a movie character who whistles heedlessly as a five-eyed space alien sneaks up behind him, I was about to be enveloped by ominous forces. The field of psychology was, along with the rest of the social sciences, about to be revolutionized—to have its foundational assumptions dynamited out from under it. Indeed, although the material world ultimately survived 1975, the conceptual world of the traditional social sciences did not. Unbeknownst to me, I was about to fall in with a band of radical scientific insurgents.

My undoing started just a few days before my graduate comprehensive examinations, at which time my learned committee members would ask me to demonstrate encyclopedic knowledge of the research and theory in the field of social psychology. I should therefore have been diligently studying the results of classic experiments testing Leon Festinger's theory of cognitive dissonance, Fritz Heider's theory of attitudes and cognitive balance, or Kurt Lewin's theory of group dynamics. But whenever I have a daunting amount of work to do, I suddenly develop an intense interest in anything unrelated to the task at hand. It was in this self-handicapping spirit that I drifted into the campus bookstore to browse around. My eye was drawn to a book called *Primate Behavior and the Emergence of Human Culture* by the anthropologist Jane Lancaster. This particular volume seemed comfortably outside the domain of experimental social psychology, so in my work-avoidance mode, I felt compelled to buy it, take it home, and read it through that very afternoon.

Lancaster's book had, as I'd expected, very little to do with the questions my social psychology professors were to ask me during my comprehensive exams. But it had everything to do with the questions they should have asked. The field of social psychology was, and is, concerned with many of the things people worry about every day: romantic love, aggression, prejudice, persuasion, and obedience to authority. Despite the breadth of topics, the scope of theory in the field was rather narrow at the time. When I entered the graduate program at Arizona State University, two of my professors had independently, and rather proudly, informed me that social psychology was a "minitheory" discipline. And sure enough, my reading of texts in the field had revealed a scattered disarray of unconnected, miniature theories, each designed to explain just a small facet of social behavior: One addressed frustration-induced aggression. Another dealt with interpersonal attraction between people with similar attitudes. Still another tried to explain responses to one-sided versus two-sided arguments. And there were many, many more such theories, mostly distinguished by their lack of connection to one another.

Social psychologists at the time prided themselves on being not only theoretically constricted but empirically narrow as well—studying anorexically thin slices of thought and behavior. Like other experimentally oriented psychologists back then, social psychologists in 1975 self-consciously rejected the study of stable "traits" as causes of behavior and focused instead on how a person's ongoing thoughts and behaviors responded to changes in his or her immediate situation. What was meant by a person's "situation" was limited to what could be captured within the half hour that a typical psychology experiment lasts. There were reasons for these strictures: Experimental studies were designed to maximize control, and theoretical restraint was supposed to cut down on rampant speculation about unobservable events inside the head or body. But to a curious young student interested in the roots of human behavior, those constraints seemed like the compulsions of the drunk who had lost his keys in a dark alley but was carefully searching for them under the streetlamp, where the light was better.

In this context, I took an almost guilty delight in glimpsing the very broad theoretical perspective suggested in Lancaster's book. It was the intellectual equivalent of what I had felt when I stumbled across an erotic magazine as a young boy in Catholic school: Here is something they probably do not want me looking at, but it sure is

hard to resist. Instead of a narrow focus on the way very specific artificial laboratory situations alter very specific aspects of social behavior in the members of our particular culture, Lancaster's evolutionary
perspective offered the tantalizing suggestion that we ought to erase
the lines between psychology, biology, and anthropology and instead
consider how all these vast subjects might fit together.

Exhilarated by the profound implications of this approach, I began
raving about Lancaster's book to anyone who would listen. Some of
my fellow graduate students and my faculty advisers just gave me an
uncomfortable smile, as if I were earnestly explaining why I had just
joined a cult. But Ed Sadalla, a new assistant professor who had recently joined the faculty, nodded knowingly. Sadalla had not seen Lancaster's book, but he had recently picked up another volume,
Sociobiology. Sadalla suggested that *Sociobiology*, which had been written by Harvard entomologist E. O. Wilson, and which dealt mostly
with the behavior of ants, lions, and other nonhuman animals, offered
a treasure chest of untested hypotheses about human behavior.

In fact, Sadalla already had a hypothesis about social dominance to
test. As part of a process that Charles Darwin had called sexual selection, females in many animal species carefully choose the males
they mate with, whereas males tend to be less selective. This, in many
cases, means that females choose the most socially dominant males,
and Sadalla was interested in seeing whether women's interest in men
was likewise influenced by their social dominance. I will discuss these
ideas in detail later. For now, I will simply say that Sadalla and I, along
with Beth Vershure, ran a series of studies suggesting that, leaders of
a new cult or no, the primatologist Jane Lancaster and the entomologist E. O. Wilson were on to something, and it was something with
powerful implications for human psychology.

Not everyone agreed. Although our findings on dominance and
attractiveness were clear and reliable, it took us over a decade to get
them published. Unbeknownst to us at the time, the armies of polit-

ical correctness were poised to sweep through academia with the combined energies of Mao Zedong's cultural revolutionaries and George Orwell's *1984* antisex crusaders. Sadalla, Vershure, and I thought we were simply applying evolutionary concepts elucidated in animals to humans, but when we tried to publish our findings, we learned we were really committing thought crimes. As one critical reviewer of our first submitted paper put it, "As a feminist and a scholar, I feel duty-bound to protect the unwary journal readership from this type of inherently sexist thinking." So dangerous were our findings that even other research scientists should be protected from them! It seems I really had been reading intellectual pornography, and Sister Katherine Mary had found me out.

The academic tumult surrounding sociobiology, epitomized by the vicious attacks on E. O. Wilson, the author of *Sociobiology*, is by now fairly well-worn academic gossip. To a young researcher experiencing it firsthand, it was a very personal battle, and it drew me into a war of ideas that would change the face of modern science. In the end, the academic controversy was often illuminating, as challenges encourage new research. And it was often fun, sometimes in ways almost embarrassing to admit. The opponents of evolutionary psychology, some of them distinguished professors at major universities, have often been so arrogant in their dismissals that they've made us look good when the actual data pronounce them wrong. In retrospect, 1975 may not have been the apocalypse for the world of traditional social science—there are still holdouts who refuse to accept that the last thirty-five years have happened—but it certainly was Darwin's second coming, and the consequences continue to unfold.

The Importance of Being Earnest

Oscar Wilde, although he had never heard of evolutionary psychology, did write the perfect slogan for the discipline: "We are all in the gutter, but some of us are looking at the stars." Part of the reason evolutionary psychologists have often upset proper academics is that we have had an inclination to root around in the gutters. A few years back, when I told my colleague David Funder that I was doing a study of homicidal fantasies, he simply rolled his eyes. Funder observed that the *modus operandi* for an evolutionary psychologist seemed to be this: Choose a topic that is normally avoided in polite conversation and shine a spotlight on it. When I thought about it, I realized this was not a completely unfair assessment. But we do not pick such topics just because they are what sells in the tabloids. Instead, we study unsavory topics (as well as nicer ones) because these are the issues with which humans the world over concern themselves—who's sleeping with whom, who might stab me in the back, who might hurt my kids, and on and on. Why do so many people read the tabloids and gossip magazines like *People* and *Us*, anyway? Because they have better book reviews than the *New York Times*, or because they have rumors about which powerful man is cheating on his wife and sleeping with which Hollywood ingenue? And why have people the world over shelled out billions of hard-earned dollars a year and stood in long lines to see movies like *Gone with the Wind*, *Titanic*, *Braveheart*, and *Avatar*? I would venture to guess it is not because those movies illustrate the finer points of cinematography, but because they present vivid conflicts between the bad guys (them) and the good guys (us), brave and heroic men involved in love affairs with beautiful young women, and other topics humans have always gossiped about.

There is more to the field than just engaging topics, however. Evolutionary psychologists are also searching for an integrated conceptual paradigm to unite the social sciences with the biological sciences. Indeed, part of what irks traditional academics is the field's apparent grandiosity—we claim that the evolutionary perspective can integrate psychology, economics, political science, biology, and anthropology, and we also insist that the perspective has profound implications for applied disciplines such as law, medicine, business, and education. And we go even further, claiming that these issues have important implications, not just for academics but for everyone—from your relatives in rural Wisconsin to the members of the UN Security Council. If there is any hope of changing the world for the better, from reducing family violence to reversing overpopulation and international conflict, economists, educators, and political leaders will need to base their interventions on a sound understanding of what people are really like, not on some fairy-tale version of what we would like them to be.

The next few chapters will focus on the research my colleagues and I have conducted on simple, selfish biases, exploring topics such as sexual attraction, aggression, and prejudice. We have considered questions such as: Why are old men attracted to much younger women? Why are older women not drawn to young men in the same way? Why does a woman's commitment to her partner drop after seeing a powerful executive, regardless of whether he is good-looking or not, whereas a man's commitment is shaken by good-looking women regardless of their social status? Why are people raised by a stepfather more likely to have fantasized about killing the old man than are people raised by a natural father? In later chapters, I will talk about how this research on simple selfish biases is connected to much broader questions about economics, religion, and society. Is fundamentalist religiosity actually a mating strategy? Can we better understand why people buy Porsches by understanding why peacocks flash their tail feathers? Sometimes mundane, sometimes shocking, our work has always been aimed at answering the biggest questions of our time: questions about what makes human beings tick. In the final chapters, I will describe how these biases, though selfish and irrational at one level, are actually deeply rational at another. And I will describe how simple biases inside individual's selfish heads combine to create complex and ordered patterns at the societal level. Finally, I will consider how an understanding of those simple selfish biases might offer us some insights about how to live a more caring and connected life.

We'll begin in the gutter, though, exploring how our very natural love for loveliness can make us miserable in surprising ways.

Chapter 2

WHY *PLAYBOY* IS BAD FOR YOUR
MENTAL MECHANISMS

To a refugee from the ice and slush of New York's winters, the sun-soaked campus of Arizona State University (ASU) was paradise found. At every opportunity, I would join several other young male psychology students on the main mall, where we would enjoy the blue skies and balmy weather while discussing the week's readings. But any semblance of meaningful conversation was disrupted for a brief interval every fifty-five minutes, when it became impossible to maintain eye contact with my fellow students, much less engage in a focused discussion of the philosophical distinctions between behaviorism and phenomenology.

The mental disruption was caused by the throng of undergraduate students parading by during the fifteen-minute break between classes. What made the break especially distracting for the twenty-four-year-old me was this: A great many of the people in that crowd of students were beautiful and athletic young women dressed as if they were on their way to audition for the *Sports Illustrated* swimsuit issue. It was a physiological challenge not to gasp. I remember thinking that the average woman at ASU was better looking than most of the people I had known growing up.

But as the mob thinned out, something funny happened. When classes were changing and there were several hundred people zipping by every few seconds, the crowd had seemed to be mostly fashion models, but when the flow of humans slowed to a mere dozen per minute, there seemed to be many more average-looking folks attending ASU. What happened to all the stunning women after classes started?

I began to consider various possible explanations of the disappearing beauties: Maybe the beautiful women attended lectures more faithfully or rushed right to the library, whereas average-looking people cut classes and spent more time drifting aimlessly around the campus mall. But that did not seem likely. Instead, I began to suspect that something else was going on, that perhaps my friends and I had been biasing our estimates of the beauty ratio at ASU. I speculated as follows: When a man's eyes scan a large crowd, they will fixate on the most physically attractive woman. When she passes, he scans the next two or three hundred people, and his eye shifts to the next beauty, who, although statistically unrepresentative, is nevertheless irresistibly eye-catching. But when the river of people shrinks to a small stream, I reckoned, you look at every individual and the mind computes a less biased average. The new mental calculation is that the average person in the smaller crowd looks just like that: an average person. That seemed to me like a better explanation, but hypotheses are a dime a dozen, and it would take two decades and some sophisticated experimental equipment before I was able to test the idea.

Regardless of any bias inherent in my cognitive estimates, though, I was fairly sure that there were more beautiful women at Arizona State than in New York. Hence I was a bit dumbfounded when my neighbor Dave observed, "There are no truly good-looking women at ASU." Dave, like me, was a recent immigrant from New York, so it did not seem likely that he and I had arrived in Arizona with grossly different expectations about what an average-looking human female should look like. And Dave's higher standard did not seem to be caused by any unique need to shoo fashion models away from his door. He was a fairly regular-looking guy, often lamenting his lack of a date for the next weekend. Why was Dave so picky? I got one possible clue when he had a party at his house and I caught a glimpse of his interior-decoration plan: Dave had wallpapered his apartment bedroom with *Playboy* centerfolds.

Fleeting Glances and Forgettable Faces

Fast-forward thirty years. It is 2002, and my research team has just received a big government grant enabling us to purchase a delightful scientific toy: a state-of-the-art eye-tracker. An eye-tracker does not allow us to read someone's mind, but it certainly gives us a better idea of what movie is playing in there. A truism of cognitive psychology is that attention is selective—that is, unless you are in a dark, sound-proofed room with your body wrapped in cotton balls, you cannot possibly pay attention to everything in your immediate environment; you would just be overwhelmed if you tried. Even sitting here quietly at my desk, I have hundreds of things in my field of view: *to the left*, glasses, a wallet, a cell phone, a coffee cup, a paperback copy of Greg Mortenson's *Three Cups of Tea*, a checkbook, a stapler, an empty plastic bag, a sideways photo of my son Liam in a dentist's chair, a pile of dust-covered zip-disks; *above the screen, Merriam-Webster's Collegiate Dictionary*, the *Random House Roget's Thesaurus*, the *Oxford Dictionary of Quotations*, and a few other assorted reference books; *to the right*, a pencil sharpener, a printer, a cylindrical container half-full of recordable CDs, a mouse (the Microsoft version), a mouse pad, and a jumble of wires; *just below the screen*, a pile of punch cards entitling me to free coffee at Gold Bar Espresso, free gelato at Angel Sweet, two passes to the Phoenix rock gym, and two human hands typing these words on a keyboard (itself composed of over a hundred keys, many emblazoned with multiple symbols, such as @, FN, ~, ALT, `, >, &, and %). That is just a partial list of what is right in front of me, and if I turn my head I notice hundreds of other objects cluttering the room. No wonder I can never find my keys!

Now imagine a student sitting on a crowded campus, with a much wider field of view, lots of people passing by in different directions, wearing a variety of colors of shirts, shorts, and shoes, some tall, some short, some with long curly red hair, some with short straight black hair, some wearing hats, various kinds of bright earrings flashing, tattoos here, political buttons there. What if this overwhelmed observer tried to pay attention to every person walking by and to everything each passerby was wearing and to all their hand-movements and to all their conversations? He or she simply could not do it, even for a few seconds. As William James observed over a century ago, the world is a "booming, buzzing confusion," made tolerable only by our capacity to ignore almost everything out there.

But our eye-tracker lets us zoom in to see exactly what, in a passing crowd of people, would catch our subjects' eye. In our laboratory experiments, we made the task a bit more manageable than if we had just sent the subjects out to the booming, buzzing confusion of the campus mall. The student observer saw only a tiny crowd of six or ten faces, frozen in his or her visual field for a short while before the next small crowd came along. Later we asked the students to identify whether or not they had seen a particular face. Even after viewing these scaled-down, slow-motion versions of real crowds, subjects were not very good at recalling whom they did or did not see. But there were some people who were easier to remember than others.

Men looking at our simulated crowds lingered almost twice as long on beautiful women as on average-looking women. When we showed them groups of photos later, the men were especially accurate at saying whether they did or did not see a particular pretty woman. When it comes to crowds of men, on the other hand, men did not gaze any longer at the George Clooney types than at the Joe Schmoes. And later on, guys were not especially good at picking handsome men out of a lineup. These findings all fit well with a traditional assumption about attention and memory: The more you attend to someone or something, the more likely you are to remember that person or thing later. But women violated that assumption in an interesting way.

The female subjects in our study, like the male ones, spent more time looking at beautiful women, and they were also good at remembering whether or not they had seen a particular female beauty before. Unlike men, though, they looked selectively at the handsome George Clooney types when presented with a crowd of men. That part was not too surprising, but what happened later was: Women were unable to remember those good-looking guys they had been staring at. This was unexpected, given that there is usually a simple linear connection between attention and memory—the more you look at someone, the more you remember them.

In later research conducted with Vaughn Becker, Jon Maner, and Steve Guerin, we asked people to play a version of the game Concentration (aka the Memory Game). In our version of the game, research participants have to uncover and match pairs of faces from a large array. Everyone was good at remembering the location of good-looking women. Sometimes we changed the normal rules of the game, though, and flashed up all the faces very briefly before asking subjects to make their matches. If we did that, women matched handsome men on the first test trial, again suggesting that good-looking guys catch women's eyes. But over trials, the advantage for good-looking men disappeared completely. Handsome men, although they grab women's attention, seem to be ejected from downstream mental processing.

This research seemed to corroborate my suspicion that men's biased attentional processes might mislead them into overestimating the ratio of attractive to average-looking women in crowds. This conclusion was directly supported by other research I conducted with Becker and Maner, now joined by my colleague Steve Neuberg and our students Andy Delton, Brian Hofer, and Chris Wilbur. In that research, we showed crowds of either women or men to our subjects. Some of the photos were of good-looking people, and others were average-looking folks (as judged by the participants in another study). Our research subjects looked at these crowds under one of two sets of conditions: Sometimes, they saw the group of faces all at once for only four seconds. At other times, they got to look at the crowd for a longer period, or they saw each face one at a time. The brief all-at-once exposure was analogous to standing on the ASU mall while classes are changing—there were too many faces to process them all. The longer exposures and the one-at-a-time presentations were more like watching the smaller trickle of people walking by between classes: They gave the mind sufficient time to reckon with the full sample.

When we strained our subjects' attentional capacities, we found exactly what I had suspected several decades before: Men overestimated the number of beautiful women (though their estimates of handsome men were unaffected). Female subjects also overestimated the frequency of gorgeous women in the rapidly presented crowds, but they did not overestimate the frequency of handsome men. The whole body of findings points to a simple conclusion about beautiful women: They capture everyone's attention and monopolize downstream cognitive processes. The conclusion about handsome men is different: They grab women's eyes but do not hold their minds; good-looking guys quickly get washed out of the stream of mental processing. This discrepancy is consistent with men's and women's different mating strategies; women are more selective and less interested in casual affairs with strangers. I will discuss the reasons for those differences in later chapters. But for now let's return to my friend Dave's *Playboy* wallpapering project.

Contrast Effects: The Trouble with the
Playboy Wallpapering Scheme

Men's default mental processes create an imaginary world with an overabundance of women who look like Halle Berry, Kate Hudson, Jennifer Lopez, and Beyoncé Knowles. On the one hand, men could take this as a beneficial side effect of overpopulation: The ensuing cognitive overload skews the world in a more beautiful direction. For women, who also mentally overestimate the number of beautiful women but not of handsome men, it is hard to see that same bright side. Other research, which I conducted with Sara Gutierres, suggests that an overdose of beauty might have ill effects for both sexes, albeit different ones for women than for men.

Around the time my *Playboy*-wallpapering friend denounced the attractiveness of ASU's female population, I was taking a course called Sensation and Perception. Perception researchers love to uncover mental illusions and errors of judgment, and one of the most robust of these is a phenomenon called the contrast effect. You can demonstrate a contrast effect on yourself by lining up three buckets full of water. Fill the one on the left with ice water and the one on the right with hot water (not so hot you cannot put your hand in, though). The water in the middle bucket should be at room temperature. Dunk your left hand in the ice water and your right in the hot water for a minute or so. Then take them both out and plunge them into the room-temperature water. What happens is a bit of bilateral mental dissonance. Your brain will receive conflicting messages: The neurons originating in the left hand will tell your brain that the water is hot, whereas the right hand neurons will signal that the same water is cold. According to a theory developed in 1947 by Harry Helson, we make psychophysical judgments by comparing any new form of stimulation to our adaptation level—an expectation of what is perceptually normal, based on our past experiences, especially the recent past. What registers as hot or cold, heavy or light, salty or sweet, depends to a large extent on what you have recently been exposed to.

I suspected that sensory adaptation processes might apply to judgments of beauty, and I set out to test the idea, working with Sara Gutierres, who was at the time a highly motivated undergraduate student but who later became my colleague and coauthor on a number of studies related to this idea. In our first study, Sara and I asked people to judge an average-looking woman after being exposed to one of two series of other women. Half the participants judged the target woman after seeing a series of unusually beautiful women; the other half judged her after seeing a series of average-looking women. As in the case of exposure to extremes of water temperature, exposure to extremes of physical appearance affected people's judgments of what was average. As we had predicted, an average-looking woman was judged significantly uglier than normal if the subjects had just been gazing at a series of beauties.

In a later study with Laurie Goldberg, Gutierres and I tried to identify whether these same processes affect judgments of people we know and love (or know and might choose to date). The cover story for that study was that we were conducting research on "community standards of aesthetic judgment." We told participants that there was an ongoing controversy about what was artistic and what was in poor taste, so we wanted to get some opinions from a random sample of students.

Subjects in the control group first judged the artistic merit of abstract paintings such as Josef Albers's *Homage to the Square*. The men in the experimental group saw centerfolds from *Playboy* and *Penthouse*; the women saw handsome naked men from *Playgirl*. After they had looked at either paintings or centerfolds, we asked our participants to rate their feelings about their current relationship partners. Again, there was a cover story—that psychologists were divided on whether being in a relationship opened people up to new aesthetic experiences or made them less open to novelty. To test which side was right, we told them, we needed to know about the extent to which our participants were in committed relationships. It turns out that their reported level of commitment depended on whether they had seen centerfolds. Again, we found an interesting sex difference: Men who had viewed the centerfolds rated themselves as less in love with their partners; women's judgments of their partners were not so easily swayed.

Generally speaking, then, exposure to beautiful women changes people's adaptation level for what they consider beautiful. The harmful side effect for guys like my neighbor Dave is this: Real women, the kind he could have dated, do not look as attractive once the mind has been calibrated to assume that centerfolds are normal. And for guys in relationships, exposure to beautiful photos undermines their feelings about the real flesh-and-blood women with whom their lives are actually intertwined.

Are women just more deeply in love? That women did not suffer the same ill effect—that handsome centerfold men did not undermine women's feelings about their partners—could be just more evidence for the well-supported hypothesis that men are jerks. But it is possible, as our colleague Norbert Schwarz suggested, that although women do not compare their mates to muscular male models, they might instead contrast them unfavorably to men who have high status.

To test this possibility, Steve Neuberg, Kristin Zierk, Jacquie Krones, and I asked students to evaluate several singles' profiles that were, we told them, part of a new program ASU was setting up to help lonely new students from out of town connect with potential relationship partners. If you were a man in this experiment, you would have seen several profiles of women, with photographs preselected to be highly beautiful or average looking. A woman in the experiment would have seen either a bunch of handsome men or average-looking men. Alongside the pictures, you would also see personality ratings. Half the time, you would be led to believe that a team of psychologists had rated the people you saw as generally high in "dominance/ascendance"—go-getters who had scored high or very high on leadership potential. The other half of the time, you would see people rated by the psychologists as submissive, follower types who were low on leadership potential. After seeing the profiles, we would ask you some questions about your own relationship history, and to make it a bit more engaging, we would inform you that if you were available and interested, you would be given a chance to sign up for the "Singles at ASU" program.

How would you have responded? The answer depends on whether you are a man or a woman. As in the earlier study, we found that beauty undermined men's commitment to their partners. Men rated themselves as least committed to their partners after they had just seen a series of beautiful and submissive women who were presumably available and on the market. Seeing a series of handsome men, on the other hand, did not substantially influence women's commitment. But if you are a woman, do not start feeling smug just yet. Seeing a series of socially dominant men undermined women's commitment, just as seeing attractive women had done to men's.

Comparing Ourselves to Starlets and Moguls

Does seeing beautiful women or successful men change how we feel about ourselves? In a follow-up study, Sara Gutierres, Jennifer Partch, and I again exposed people to profiles of other folks who, we told them, had signed up for the "Singles at ASU" program. In this study, though, we had the subjects look at profiles of people of their own sex.

Each profile included a name, a short list of hobbies and interests, and the person's "most notable accomplishment." There were two versions of each profile, depicting similar interests and activities but varying in whether the person described himself or herself as high or low in social dominance. For example, here is what "Carl Powers" had to say about himself in the high-dominance condition:

> I think that I have plenty of friends because people can count on me and I enjoy a good time. I like to plan new adventures for my friends and myself. I work out 5 days a week and teach trampoline to kids at the Y on the other days. I like to be with people and I often end up as group leader when someone needs to take charge. I like being in leadership positions; it comes easily to me, and I get to meet a lot of people that way. I'm told that I'm a natural at delegating responsibility to others. I was really pleased to be chosen editor of the campus newspaper at the University of Washington before I transferred to ASU. I've already published two short pieces in *Runner's World* magazine, both of them about the qualities that it takes to achieve excellence. I try to practice what I preach and that's probably what accounts for my success so far.

In the low-dominance condition, Carl was not such a go-getter, but instead was rather meek:

> I think that I have plenty of friends because people can count on me and I enjoy a good time. I'm usually willing to go along with whatever adventures my friends plan for us. I try to go to the gym frequently and help out with the children's trampoline program at the Y on other days. I like to be with people and I'm not too proud to run errands or help in anything that needs to be done. I don't like being in leadership positions. It doesn't come easily to me, and it gets in the way of getting to know people, but I'm pretty good at carrying out the responsibilities that get delegated to me. I was really pleased to be chosen most helpful employee of the campus newspaper at the University of Washington before I transferred to

ASU. I've been writing a couple of short pieces I'd like to get published in a magazine, both of them about the qualities that it takes to be content with yourself. I try to practice what I preach and that's probably what accounts for my own contentment.

Some people saw eight high-dominance profiles, designed to give the impression that there were a lot of successful and hard-driving members of their own sex out there competing with them for mates. Others saw eight low-dominance profiles, suggesting a less imposing set of competitors. Women saw identical profiles, except with female names (e.g., "Amy Powers" instead of "Carl Powers"). Each profile had a photograph attached, so that some subjects saw eight highly attractive members of their own sex, whereas others saw eight average-looking potential competitors.

Looking at highly attractive or highly successful members of their own sex did not change people's self-ratings of how attractive or socially dominant they thought they were. But it did change their guesses about how they would be rated by other people. And the changes were flipped from those we found in people's ratings of their own mates: Men who had been bombarded with socially dominant men felt that they themselves were less desirable as marriage partners; women who had been bombarded with physically attractive women downgraded their own marriage prospects.

Candy for the Mind: Watch Your Portion Size

So our research suggested that if you are a man, exposing yourself to too much beauty can undermine your feelings about your partner; if you are a woman, on the other hand, exposing yourself to too many go-getters can chip away at your commitment. But in the modern world, starlets and moguls seem to be everywhere. In Hollywood's market-driven attempt to give the people what they want, movies and television shows (from *Gone with the Wind* to *Who Wants to Marry a Multi-Millionaire?* and *Mad Men*) tend to be populated with more beautiful young women and rich powerful men than you will ever see at the local shopping mall. To the extent that you expose yourself to movies and television, then, it might be bad for your partner. And probably for you, too.

Open up a popular magazine, turn on your television set, or go out to the movie theater, and you will enter a world full of beautiful women and powerful men. Our research suggests that unless you already live in Hollywood, your innocent attempts to entertain yourself may be making the people in your real life pale in comparison and may be undercutting your own confidence. Should you write your senator and lobby for legislation demanding equal media representation for your average-looking average-accomplishing neighbors? Or better yet, why not demand that the media be filled with grossly unattractive and unaccomplished losers? Then, when we turned on our television sets, we would get to feel better about ourselves and our romantic partners—until, of course, we discovered the black market of beautiful-jet-setter videos that would spring up to fill the void.

Why are we so attracted to these images? My guess is this: Our minds are designed to look for the beautiful and the powerful because our ancestors either picked the local beauties and bigwigs as mates or competed with them to get mates. It paid to be aware of the opportunities and threats out there. Of course, our ancestors lived in worlds without television sets, movies, or photographs, so they only saw real people, and their mental mechanisms did the jobs they were supposed to do. Now those mechanisms are overwhelmed. In a sense, the images from Hollywood and Madison Avenue are analogous to the flavors of Ben & Jerry's ice creams. The tasty flavors and images tap into mechanisms that were designed to promote survival and reproduction in a much different world. Consume too much, though, and it may be harmful to your health.

So what's a mortal to do? Are we helpless in the face of our evolved mechanisms, which may lead us astray without our conscious awareness? Not completely. People who understand the dangers of overabundant fats and sugars can control their diets. People who understand the dangers of an overabundant diet of mass-media images can stop gorging on *Playboy*, *People*, *Sex and the City*, or *Dancing with the Stars*. After doing all this research, I do not buy *Playboy*, and I almost never turn on a television set. As a consequence, I have more time to go out for a bike ride or to read a book (the kind without pictures, unless I'm reading Dr. Seuss to my younger son). Whether I'm happier, I don't know, but at least I'm not comparing my successes to Donald Trump's or my wife to the airbrushed charms of Miss February.

In this chapter, we have considered one potential cost of not understanding our natural inclinations, of not seeing how our visual affinity for beauty can, like our natural taste for ice cream, have insidious effects over the long haul. Now let us shift our attention to something much more immediately dangerous—natural inclinations that, if given free rein, could very quickly land us in prison (as almost happened in my own case) or in the morgue.

Chapter 3

HOMICIDAL FANTASIES

S teve Lowry was one of the fellows who used to sit around with me
on the ASU mall discussing philosophical issues and overdosing
on beauty. At first glance, Steve and I appeared rather similar; we were
both tall, long-haired, bell-bottomed white male graduate students in
clinical psychology, we both loved to play the guitar, and we both enjoyed discussing topics like phenomenology and existentialism late
into the night. But we actually came from radically different cultural
backgrounds. Steve had grown up in an upper-middle-class suburb
in Ohio and claimed he had never been in a fistfight in his life. Having grown up in a New York neighborhood with more than its fair
share of lower-class Irish and Italian hooligans, I had a hard time be-
lieving him. There were periods during my childhood when I had a
fight every day. And I was surrounded by people who were quite a bit
tougher than me and had a father in prison and plenty of friends and
relatives who would eventually end up there as well.

My stepfather Bob had middle-class aspirations, and I can thank
him for getting me out of Queens and away from a crowd of ruffi-
ans. But Bob had been raised in the same neighborhood, so even
after we moved out to the land of Ozzie and Harriet, his idea of
parental advice still involved the occasional punch in the jaw. Bob
was a proud member of the National Rifle Association and had a

gun rack hanging prominently in the kitchen; when he got drunk, he would lose his otherwise pleasant disposition and threaten to shoot my brother and me if we tried to intervene in one of his battles with our mother. (She also had a taste for liquor, as well as a special knack for egging Bob on.)

One evening my stepfather was especially out of control and came at me with fists flying, shouting threats against my life. In a scene from the movies, I managed to land the best punch of my pugilistic career—a square hit on his jaw that sent him flying across the room, where he fell unconscious on the floor. My brother looked at me, pointed to the guns Bob had just been threatening us with, and asked, "Should we kill him?" I actually had to think about it for a minute before I said, without complete resolve, "Nah, we'd better not."

So with this background, I was no stranger to thoughts of homicide. Indeed, when my colleague Norbert Schwarz expressed doubt at my assumption that everyone had homicidal fantasies, I thought he was putting me on. When I surveyed the other colleagues with whom Norbert and I were eating lunch, though, they were split: Some claimed they'd never had a homicidal fantasy, but the others accused them of denial. For experimental psychologists, this kind of disagreement means there is an interesting hypothesis waiting to be tested. So I decided to do a more systematic investigation, working with my graduate student Virgil Sheets, another midwestern boy, but one who hailed from a rougher town than Steve Lowry's.

Everyday Murderous Thoughts

A few days after the debate with my colleagues, I gave a lecture on aggression. During the class, I asked my students to fill out a brief questionnaire about their own violent fantasies. To overcome the normal tendency to hide antisocial inclinations, I explained that even normal people occasionally have angry and violent thoughts about others, which might range from thoughts about "telling someone off" or hitting someone to thoughts about killing another person. I then asked the students to describe a recent time they had had thoughts about killing someone or, if they had never had such thoughts, to tell us about the most violent thoughts they had had.

Over the next year, Virgil Sheets and I asked these questions of 760 ASU students. The results were clear: The majority of those smiling, well-adjusted, all-American students were willing to admit to having had homicidal fantasies. In fact, 76 percent of the men reported such fantasies, firmly in line with my more pessimistic expectations. I was somewhat surprised, however, to find that 62 percent of the so-called gentler sex had also contemplated murder at least once.

Is there something especially violent about students in Arizona? Probably not. When David Buss and Josh Duntley later surveyed a sample of students at the University of Texas, they found similarly high percentages of men (79 percent) and women (58 percent) admitting to homicidal fantasies. However high these numbers might seem, my guess is that they underestimate how many normal people have homicidal fantasies. Social psychologists have found time and again that people are motivated to say what they think is most socially desirable and that we all tend to be selectively forgetful of evidence that we are not good little girls and boys. As in the case of Alfred Kinsey's famous data on self-reports of masturbation, it is probably safe to assume that the actual incidence of homicidal fantasies is at least as high as what people are willing to publicly admit.

Whom do people think about killing? Both sexes were inclined to target men; 85 percent of men's fantasies and 65 percent of women's involved killing a man. This part was not too surprising, given actual homicide statistics, which show men much more likely to be homicide victims. There were several interesting sex differences in the homicidal fantasies, though. We found that 59 percent of men (as compared to 33 percent of women) had fantasized about killing a total stranger. Indeed, for 33 percent of men (as compared to 10 percent of women), their most recent homicidal fantasies involved a stranger. Women outnumbered men only in the category of fantasies about romantic partners: Among women, 27 percent, as compared to 7 percent of the men, reported that their most recent fantasy was about a romantic partner. This was due not so much to the fact that men were innocent of ever contemplating extreme relationship violence—they are not—as to the fact that men have more frequent fantasies about killing people in other categories. And of course, women's relationship partners are men, who seem to inspire more homicidal fantasies in everyone they meet.

Despite those differences, I was struck by the unexpected similarity, by the fact that women's likelihood of having had a homicidal fantasy was so close to men's. I had expected that women would be much less likely to have homicidal fantasies. Consider the data on actual homicides, where the sex discrepancies are whopping. Every year in the United States, men commit approximately 90 percent of the murders. And that discrepancy is in no way peculiar to the United States. When Martin Daly and Margo Wilson reviewed data from other societies and from other periods in history, they found this same wide gender gap all around the world and throughout history. They found that homicides were overwhelmingly a male affair in modern Canada, Australia, and Scotland; in Miami during the 1920s; and in Birmingham, Alabama, during the 1940s. The same is true for remote tribal groups such as the Tzeltal Mayans. And Daly and Wilson were able to dig up data on homicides that had been perpetrated between the years 1296 and 1398 in Oxford, England. Again, when medieval Brits killed medieval Brits, it was overwhelmingly men wielding the swords and axes.

There are at least three possible explanations for the fact that women are less likely to act on their homicidal fantasies. First, female fantasies tend to be more fleeting than those of males; 65 percent of women who experienced homicidal fantasies said that they only lasted for seconds or minutes, but men were more likely to report that their fantasies lasted for hours, days, or even weeks. Men's fantasies were also more likely to involve devilish details and were more likely to engage the men's planning capacities. Consider the following fantasy reported by one guy:

> I wanted to kill my old girlfriend. She lives in Albuquerque and I was just wondering if I could get away with it. I thought about the airline ticket and how I might set up an alibi. I also thought about how I would kill her in order to make it look like a robbery. I actually thought about it for about a week and never did come up with anything.

Women's fantasies were less detailed and were often as simple as "I just wanted his car to run off a cliff." It takes more than such fleeting ill wishes for someone to end up dead.

A second possible explanation of the fantasy-reality sex gap is that females have stronger inhibitions that block their violent impulses. Aggression researchers Kaj Björkqvist, Kirsti Lagerspetz, and Ari Kaukiainen explain some of the sex differences in adolescents' violent behavior in terms of what they call the "effect/danger ratio"— or the person's assessment of the likely beneficial effect of aggressiveness, balanced against the likely dangers. So, for example, when my KO punch temporarily halted my stepfather's drunken rage, my brother and I were petrified that he would come gunning for us, and in fact we hid out for several days afterward. This ratio of dangers to benefits is generally even more unfavorable for a woman aggressing against a man (the most common target of women's homicidal fantasies).

Throughout most of human evolutionary history, a woman aggressing against a man could face severe danger if the man retaliated. Before the advent of the gun, the average male could probably fend off most attacks from a female and was likely to strike back in anger if he had been wounded. It would thus have been adaptive for females to quickly inhibit extremely violent impulses. Indeed, Daly and Wilson note that the typical woman who kills her husband or boyfriend most often does so defensively—to protect herself against an abusive man who she fears will someday murder her if she does not strike first.

Aggressing to Impress

The third possible explanation for the sex difference in actual homicides is linked to a surprising motive for much violent behavior: the tendency to act aggressively to impress others. This motive is rare in women but prominent in men, and the proclivity to show off their violent tendencies may explain why men are more willing to translate homicidal thoughts into actions.

Consider the famous incident in which Al Capone invited Albert Anselmi, John Scalise, and Joseph Giunta to a banquet in their honor. After wining and dining his three fellow mobsters, Capone reputedly had his henchmen tie them to their chairs. He then picked up a baseball bat, and in front of the other dinner guests, personally proceeded to beat each of the three men to death. As the most powerful man in Chicago, with politicians and police officers as well as hordes of other mobsters on his payroll, Capone usually had his underlings do the dirty work. Why, then, would he commit a triple murder right in front of a room full of witnesses?

The answer is that Capone had learned that the three were plotting against him, hoping to advance their own careers. As a powerful mafioso, he was expected to punish such disloyalty with death.

In this case, he hoped not only to eliminate these potential competitors but also to send a powerful message to his other business colleagues.

The stakes were high in Capone's world during the days of Prohibition—a life-and-death game for control of Chicago's multimillion-dollar alcohol-running territories—but the sad truth is that men will fight to the death even when the stakes are considerably lower. In his classic study of homicides in Philadelphia, Marvin Wolfgang categorized 37 percent of the causes as "trivial altercations" over relatively petty issues, such as an insult, a curse, or one person bumping into another. As one Dallas homicide detective put it:

> Murders result from little ol' arguments about nothing at all. Tempers flare. A fight starts, and somebody gets stabbed or shot. I've worked on cases where the principals had been arguing over a 10 cent record on a jukebox, or over a one-dollar gambling debt from a dice game.

It is not that women are not sensitive to social put-downs; they are. But only men are driven to kill over them. And they do it with surprising frequency. In fact, Wolfgang found that trivial altercations were the most common motives for men's murders—more important than disputes over money, property, or infidelity. Why kill over such small stakes? After an extensive examination of police reports of homicides, Wilson and Daly suggested that the stakes were actually not trivial at all. Instead, the trigger for extreme violence is not the content of what one man says to another, but how he says it and what that tone implies. When one man openly insults another in public, regardless of the trigger for the insult, the insulted man's status is being challenged. And when a man loses status in the eyes of other men, Wilson and Daly argued, his ability to attract women also takes a hit.

The link between a man's status and his value on the mating market connects to two of the most important principles in evolutionary biology: sexual selection and differential parental investment. According to the principle of *differential parental investment*, when one sex (usually the female) invests more in the offspring, members of that sex will be more careful about mating. As a consequence, members of the other sex (usually the male) will need to compete to be chosen. Consistent with this principle, human females, because they can become pregnant, have more to lose from a rash mating decision. Hence women tend to take more care in choosing the men with whom they mate. The process through which males are chosen is known as *sexual selection*. To win the attentions of selective females, male animals can do one of several things. They can display positive characteristics, as when a peacock displays his extravagant tail. They can find and control a resource-rich territory. Or they can beat out the competition directly—by fighting their way to the top of the local dominance hierarchy. Whether the game is defending a territory or winning a place at the top of the hierarchy, it helps to be larger and more aggressive.

And this process can indeed work in the other direction. Consider a group of interesting little shorebirds called phalaropes. In the phalaropes' case, males are the ones who brood and rear the chicks, and so they are choosy about the females with whom they will mate. As expected from the general principles of sexual selection, the male phalaropes are small and drab, and the females, who do the courting, are larger and more aggressive.

So phalaropes are the exception that proves the rule: The sex that invests more in the offspring is choosier about mating, and the other sex will compete to be chosen. In this equation, aggression is a by-product of that competition. Returning to the sex difference in homicides, it is human females who tend to invest more in the offspring, so males need to compete to be chosen. Sometimes the competition becomes deadly.

Experimenting with Status-Linked Violence

Male aggressiveness is not a constant: It ebbs and flows according to several factors. For example, in many species, it increases just before the mating season, when territories and females are being contested. In humans, boys boost their dominance displays after they hit puberty, when successful competitiveness, such as being a star athlete, translates into popularity with the opposite sex. And men are most dangerous in their late teens and twenties, when their testosterone levels are highest and when they are competing most vigorously for mates. On the other side, when a man gets married, his testosterone level drops, and when his wife has a child, it drops again. There is less need to show off, and more need to stay out of potentially deadly competitions over which song is playing on the jukebox at the local bar.

Even for men who are fully on the market, violence is an expensive and dangerous route to respect—and one that, other things being equal, men would typically do well to avoid. In fact, they do generally avoid it. It is only when other paths to status are blocked that men resort to violent and antisocial behavior, as psychologists Jim Dabbs and David Rowe argued, with a great deal of evidence to back up their arguments. Rich men, even those with high testosterone levels, do not typically go around getting into fistfights. They can win more respect by making clever investments or perfecting their golf swings. Compare my friend Steve Lowry from the upper-middle-class suburbs in Ohio, who had never been in a fistfight, to the belligerent (and often bloody-nosed) hooligans I grew up with in New York. Lowry was a master of philosophical argumentation and could stand above the other middle-class guys by showing off his knowledge of Søren Kierkegaard. In my neighborhood, using the word "existentialism" in a sentence would have been less likely to elicit respect than the question "What are you, a fuckin' faggot?" So the urge to compete need not lead to violence, but depending upon the environment, upon the person's other traits, and upon his or her current life situation, it can.

Vlad Griskevicius and I decided to investigate how aggressive competitiveness might rise and fall as a function of a man's desire for status or mating. Vlad is an imposingly large and socially dominant guy who was born in the Soviet Union and spent his teenage years in urban Los Angeles. Along the way, he learned something about aggressing to impress, as well as about evolutionary theory. Vlad likes to tell a story about the way basketball player Charles Barkley responded when some local wise guy threw a glass of ice water on him in a discotheque. Barkley picked up his harasser and threw him through a plate-glass window. As he was being led out in handcuffs by the police, a reporter reputedly asked Barkley if he regretted his actions. Barkley replied, "I only regret that we were on the first floor." Along with Josh Tybur, Steve Gangestad, Elaine Perea, and Jenessa Shapiro, we set out to experimentally examine how those transient motivational states might alter people's responses to insults.

To begin, we surveyed college students about their actual life experiences with public insults and with violence. Apparently, there are plenty of people hurling public insults out there. We found that 75 percent of both men and women had had at least one experience in which another person insulted them to their face in a public setting. How they responded to those insults varied: Some people simply walked away; some responded with indirect aggression, such as bad-mouthing their aggressor to others or spreading a bit of nasty gossip; and some responded directly and aggressively, taking a swing or yelling at their insulter. Of the various options, men most often responded with direct aggression, whereas for women, indirect aggression was the most common response.

In a follow-up laboratory study, we confronted our subjects with three scenarios, designed to manipulate their motivations. One set of subjects imagined being on the last day of a vacation on a tropical is-

land and locking eyes with someone very attractive. As the story unfolds, they become more and more entranced with this person's company, and after a romantic dinner, a moonlit walk on the beach, and
increasingly intimate brushes with one another's fingertips, the two
end up passionately kissing one another. We had the second set of
subjects imagine arriving for the first day at a new job with a presti-
gious company, where they meet two other people who have also just
been hired. Their new boss informs them that after six months, one
will be fired and one will get a big promotion. Finally, we had subjects
in the control condition imagine neither love nor status but, rather,
envision a scenario in which they had searched for, and ultimately
found, a missing wallet.

After priming the different motives, we asked all the subjects to
imagine the same scenario: that they were at a party and that a class-
mate they knew spilled a drink on them and failed to apologize. Their
choices were to (1) hit the person, (2) insult the person to his or her
face, (3) push the person, or (4) get in the person's face (all forms of
direct aggression); or to (5) talk behind the person's back, (6) tell a
friend an embarrassing secret they have heard about the person, (7) try
to exclude the person from a social group, or (8) make up a lie about
the person (the latter four being forms of indirect aggression).

Whether or not a person preferred to act aggressively depended
on his or her motivational state and on whether or not he or she was
a she or a he. For men, thinking about status increased their desire to
hit, push, or get in their insulter's face—"You talkin' to me?" Thinking
about courtship and romance, on the other hand, slightly suppressed
men's inclinations to attack. For women, neither of the motives increased their desire to fight directly over an insult. However, both status and mating motivations did increase women's desire to retaliate
indirectly.

However, in certain circumstances romantic motivation can make
men more violent. In the first experiment, students had imagined

themselves at a party where members of both sexes would be present. In another experiment, we had some of the subjects imagine instead that the audience to the insult was composed entirely of the members of their own sex. When men imagined being insulted in front of other guys, both romantic and competitive motives inspired them to want to act aggressively.

So the bottom line of this series of studies is this: Either status or mating motives can lead men to want to be directly aggressive. But men seem to realize that violence itself is not sexy to women. Hence, a man in a mating frame of mind is inclined to behave himself in front of women but to be especially prone to show off his aggressive reactions if the audience is made up of other men. To the question about why men fight in bars even when there are no women to impress, the answer is that the show is in fact for the other guys—it is a gambit to hold onto one's position in the male dominance hierarchy, not to win love directly. Indeed, in many other species, the males arrive at the mating area several weeks before the females and do all their head-banging before the females arrive. The females do not need to see the fight. They merely want to know who won.

Levels of Analysis: Why There Is More Than One Why

The research I just described presumes a fundamental connection between aggression and reproduction. A reasonable person might protest, "When I got into a fight with that guy in the bar, it had nothing to do with any desire to reproduce; it was solely because the guy insulted me!" This objection is quite reasonable on one level; our conscious minds process what is going on in the immediate environment and may be completely unaware of any connections between our reactions and their underlying evolutionary roots. The disconnect has been a problem for biologists as well as for laypersons. Indeed, opposing groups of biologists talked past one another for years, having heated arguments about the causes of animal behavior, without real-izing that the other side was working at a different *level of analysis*. The famous paleontologist and science writer Stephen Jay Gould often missed this point, posing false "alternatives" to evolutionary ex-planations, and as a consequence he did not endear himself to many evolutionary biologists. To appreciate the importance of distinguish-ing different levels of analysis, consider the question of why mam-malian mothers nurse their offspring. This question can be answered at several different levels of analysis:

1. *Functional* or evolutionary explanations focus on the ultimate adaptive purposes of behavior. If we say that mammalian moth-ers nurse their young because it increases offspring survival rates, we are offering a functional explanation.

2. *Historical* evolutionary explanations focus instead on the ances-tral roots of a feature, trait, or behavior. For example, we can say that humans nurse their offspring because they have mammary glands as well as a set of associated hormones and attachment mechanisms passed down from our mammalian ancestors.

3. *Developmental* explanations focus on the lifespan events that sensitize animals to particular cues in the environment. A de-velopmental explanation would be that mothers nurse offspring because of a sequence of lifespan events involving puberty, preg-nancy, and childbirth, which combine to lead to the capacity to produce milk.

4. *Proximate* explanations focus instead on the immediate trig-gers for a given behavior—what is going on in the animal's body in response to events in the environment in the *here and now*. A proximate explanation might be that nursing occurs because an infant has begun suckling on the female's nipple, which leads to immediate hormonal changes that stimulate milk release.

Sometimes there is an obvious connection among the different levels of analysis. In the case of nursing, for example, it is easy to see the
links among the infant suckling, the development of breasts, being a
mammal, and the functional benefits of providing nutrition to the
young. But the connections among the four levels of analysis are not
always so clear. Consider the question of why birds migrate each year.
Here is a proximate explanation: Birds migrate because days are getting shorter—the immediate cue that triggers migration. But the
functional reason birds migrate has nothing at all to do with the
length of days, per se. Instead, they migrate because the locations of
the best food and the best mating sites change with the seasons. Birds
do not need to be aware of the indirect connections among day length,
seasons, survival, and mating. In fact, it is a safe bet that most animals, including humans, are completely unaware of most of those
sorts of connections. This is another point that confuses critics of evolutionary psychology. When someone says, "I have sex because it feels
good, and I do my best to avoid having children; it has nothing to do
with any motivation to propagate my genes," they are absolutely right
at the proximate level (what is going on in their heads in response to
events in their environment) but dead wrong at the level of evolutionary function. Throughout this book, we will be talking a lot about
the links between evolutionary function and proximate influences, so
it is important to understand the distinction between different levels
of analysis.

The evolutionary influences on behaviors are not directly available
to consciousness, whether for migrating birds or for humans acting
aggressively to impress other humans. But it is also important to appreciate that many proximate influences on our behavior are not available to consciousness either. For example, evolutionary researchers
have been discovering a number of connections between hormone
levels and social behaviors of various sorts. Unless you are an en-
docrinologist, you probably have no mental representation whatso-

ever of your hormones (though you may experience feeling excited, fearful, or sexually aroused, which are downstream effects of those hormones). Likewise for feelings of anger, which are linked to hormones that start flowing in situations in which it would have benefited our ancestors to act aggressively. But when you are angry, you do not think, "I am experiencing a surge of testosterone and noradrenaline, and I believe that, in the interest of enhancing my reproductive success, it would be wise to yell at this person who is challenging my status." Instead, you are thinking, "This jerk is one seriously irritating and disrespectful asshole!"

When Women Get Direct

Although women commit fewer assaults and homicides than men, it would be a mistake to conclude that all human females are harmless St. Theresa–like little flowers. There is the occasional Lizzie Borden in the mix. Indeed, the "mere" 10 percent of American homicides perpetrated by women still adds up to several thousand per year. In a review of the literature on this topic, evolutionary psychologist Anne Campbell summarized the conditions under which women will kill as "stayin' alive"—women may act violently if their own survival, or that of their offspring, is threatened.

Consider one particularly vivid historical example. In 1789, France was in chaos, gripped by a severe economic recession and widespread famine. While poor women and their children were starving, they heard rumors that their queen—a young Austrian named Marie Antoinette—was continuing to throw away the state's money on banquets, jewels, and other extravagant luxuries. One day, an angry crowd of women began marching from Paris to the royal palace at Versailles, the crowd growing larger as other women joined them along the way. By the time they arrived at the palace, there were several thousand women, wielding axes, bayonets, and pikes and crying out for bread.

When no bread appeared, they began chanting for Marie Antoinette's head. Although the women were unsuccessful at finding the queen, they did find one of her bodyguards and decapitated him. In modern times, poverty is still linked to violence in women as well as in men. In areas with high numbers of people on unemployment and welfare, and during times of acute resource shortage, women are more likely to commit violent crimes.

In an experiment designed to investigate the triggers of female violence, we asked another group of students to imagine the following scenario: You've just graduated from college and the country is entering a recession. After spending months looking for work and exhausting all your savings, you can't count on any more financial support from your friends or family. Finally, you land a job at a large company. But you discover that to keep this job, you will have to compete with two other women (or men, if you are a man). As in the status-competition story, one of you will be fired and another will have a shot at a big bonus. Thinking about losing their jobs and facing high debt was the one motivational factor that produced a substantial boost in women's inclinations to approve of direct aggression. Interestingly, a similar pattern is found among chimpanzees: Anthropologist Martin Muller found that when scarce resources or feeding territories are at stake, females chimps begin to act like the normally more aggressive males.

So under some circumstances, females will switch from their preferred strategy of indirect aggression to a directly aggressive strategy. The circumstance most likely to trigger that shift is a severe economic threat. Incidentally, although Lizzie Borden was never actually convicted of the brutal murder of her father and stepmother, one aspect of her story fits with the general story of female direct aggression. Before the murder, Lizzie and her sister, both spinsters living in their father's house and dependent on his economic support, had been having bitter arguments with the old man over his plan to divide up his valuable properties before he died, including giving away a house to relatives of their stepmother.

Why Do Men Fantasize About Killing Strangers?

I mentioned earlier that the majority of men in our survey had had at least one homicidal fantasy about a total stranger. At first glance, this is an odd phenomenon; hand-to-hand combat with a total stranger can be dangerous, and even if that anonymous fellow is an especially rude driver, he probably does not qualify for a death penalty. What makes it most puzzling is that it is hard to see the benefits of expressing one's road rage toward a rude stranger on the highway, where even the audience is composed of strangers.

One possible explanation is that strange men are automatically categorized as especially threatening. When my older son Dave was a boy, he would have nightmares about unknown men chasing him. Little Davey's nightmares about dangerous male strangers were right in line with systematic data on children's dreams collected by Michael Schredl of the Mannheim Mental Health Institute's Sleep Laboratory. Schredl found that over 50 percent of the human aggressors in boy's dreams were unfamiliar men. By contrast, none of the boys had nightmares about unfamiliar women.

So "bad guys" are usually guys, and they are often unfamiliar. In a related line of research with Vaughn Becker, Dylan Smith and I asked students either to "think of an angry face" or "think of a happy face." When people were asked to think of a happy face, the majority envisioned a woman, and it was typically a woman they knew. When they thought of an *angry* face, though, 75 percent of our participants spontaneously thought of a man. Most interestingly, that man was typically not someone they knew—so they were calling to mind not a real person with whom they had had an actual conflict but an ominous Jungian prototype—the angry strange man.

The people most likely to compete with you for status, to annoy you on an everyday basis, to bully you, or to otherwise make your life miserable are much more likely to be people you know. So why do people waste energy on feelings of antipathy toward total strangers, and why do men occasionally end up dead or in prison when they express those negative feelings toward a fellow they would otherwise never see again? In the next chapter, I will describe how that mystery can be solved by understanding the evolutionary psychology of prejudice.

Chapter 4

OUTGROUP HATRED IN THE BLINK OF AN EYE

For some of my hipper friends, 1969 was the summer of peace, love, and Woodstock. For me, though, it was the summer of learning to sing "John Jacob Jingleheimer Smith" along with a crowd of screaming five-year-olds. It was also the summer of lessons in silly human prejudices.

I had landed a job as a camp counselor at a summer camp for the children of upper-middle-class Long Islanders. The pay was terrible, and the kids were loud and spoiled, but for a college guy, the benefit package included some perks—most importantly, many of the other counselors were college females of the healthy outdoorsy type. Within a short while, I began dating one of my coworkers, a very pleasant and attractive dark-haired young woman.

Although she seemed to like me well enough, my new romantic interest never wanted me to pick her up at her house. The reason was that her grandparents, devout Jews who lived with her family, would have been mortified at her dating a goy. I had grown up in a neighborhood where non-Catholics were minorities, so I was more amused than offended by their reaction (her grandparents had lived through the Nazi years, so their distrust of Gentiles could be easily forgiven). But I was offended at my own mother's negative reaction when I brought this lovely girl into our house. My mother had been raised Catholic, schooled by the same nuns who had instructed me to "Love thy neighbor." Mom had not been to Mass for over a decade, since divorcing my shiftless Mick of a father and marrying a Protestant, so I didn't expect her to be narrow-minded. And the fact that she was a mildly liberal Democrat who had worked on John F. Kennedy's campaign also led me to expect tolerance from her. But she commented disapprovingly, "Douglas, I can't believe you're dating a Jewish girl!"

I spared the Jewish grandparents and my erstwhile Catholic mom some grief, because I started dating another one of my coworkers, a strawberry blonde whose last name was Wilhelmson.

I fell deeply in love with Ms. Wilhelmson. In fact, we soon decided we would get married. When she brought me home to meet her mother, though, I was treated to yet another round of tribalism: My future mother-in-law desperately wanted her daughter to marry a Lutheran—and not just anyone from that sect would do. She was unhappy that her son had married a German Lutheran. She wanted nothing "less" for her daughter than a Scandinavian Lutheran like themselves. Even Martin Luther himself would not have qualified. When I went to one of the family's Christmas smorgasbords, her Svenskfolk kin alternated between speaking Swedish to one another and grousing about the "Gottdamtd Puerto-Ricans, who come to this country and don't learn to speak English." After a few beers, I unwisely brought up the topic of ethnic tolerance, to which one of the Svenska guys responded, in a thick Swedish accent, that "Hitler had the right idea!" My future wife did not share their full array of Swedish Lutheran values, though, and we married anyway, in a Lutheran church, which again troubled my mother the lapsed Catholic.

A Failure to Discriminate

So I have seen people make silly distinctions all my life, between different Caucasian tribes and between different Christian sects. On the other side, though, sometimes the failure to discriminate can be just as prejudicial. Consider the case of Lenell Geter. Geter was an engineer working at a research center in Dallas. News commentators were shocked when he was handed a life sentence after being convicted of robbing a Kentucky Fried Chicken outlet. The shockingly stiff sentence was even more surprising given that there was absolutely no physical evidence linking Geter to the crime and that his coworkers had testified that he was fifty miles away at the time of the theft. There was not much of a motive, either. Why would a working engineer risk a lucrative career for a $615 stickup? But the all-white jury ignored all that, and instead trusted the testimony of the eyewitnesses, who were either white or Hispanic and who expressed strong convictions that this particular fellow was in fact the guilty party. His coworkers and the National Association for the Advancement of Colored People fought to have the evidence reconsidered, but Geter sat in prison for over a year, until police arrested another man involved in a string of similar robberies, and the confident eyewitnesses now identified the new suspect as the robber. If you look at photos of Geter and the real crook, it is surprising how different they appear. As Geter joked to my colleague Steve Neuberg, "I'm much better looking than the other guy" (and he is). But they did have a few features in common: Both were young, both were men, and (most importantly) both were black.

Geter had fallen victim to a well-known psychological bias called outgroup homogeneity. Several decades of research has revealed that most of us are a whole lot better at distinguishing among members of our own groups than among members of other groups. As with most of our habitual cognitive biases, there is an underlying functional logic to outgroup homogeneity. We usually have more experience sorting out the members of our own groups, and it is usually a lot more important for us to make distinctions among the people with whom we interact on a daily basis. When we do interact with members of outgroups, it is often at the group level rather than at the individual level (for example, if the members of an Amsterdam soccer team are taking the train from Florence to Naples, they need only distinguish the Netherlanders from the Italians). By analogy, unless you are an ornithologist, you may not know the difference between a black-capped chickadee, a boreal chickadee, and a bridled titmouse, and if someone pointed one out, you still might see just a small chattering bird.

Are there times when it might be functionally important to be able to distinguish the members of other groups? Our research team investigated this question in a series of studies headed up by Josh Ackerman and Jenessa Shapiro. We reasoned that the typical tendency to mix up outgroup members might vanish when one of them is angry. We had several reasons for our hypothesis. For one thing, it would pay to take heed when someone near you is pissed off, because he or she might attack you. Unlike a member of your own group, who is linked to you and may even be a relative, an outgroup stranger has less to lose from doing you harm. For another, anger is very personal and particular—it typically signals a threat from one specific person (the one who is angry) to another specific person (maybe you). For a third, angry expressions are fleeting, and an angry person may try to hide those feelings, even while still thinking about hitting someone. So it is good to remember exactly which person was just flashing that angry look.

To test the hypothesis that angry expressions would erase the outgroup homogeneity effect, we showed our subjects photographs of black and white men whose faces wore either obviously angry or nonthreatening, neutral expressions. To make the task more challenging, we showed our subjects each photo for only a half second while also distracting them with an abstract painting that appeared on the screen alongside the face. Afterward, we tested our subjects to see how well they could remember the faces they had seen. The test was sort of like a police lineup: Subjects had to distinguish between the photos they had seen and another set of similar faces.

For neutral faces, we found the usual outgroup homogeneity effect. Our undergraduate subjects (mostly whites and Hispanics) were better at remembering unemotional whites than unemotional blacks, and they regularly gave false positive identifications of black faces. In other words, white people were falling prey to the "seen one, you've seen them all" problem. Something different happened for the angry black faces, however. People did not homogenize those faces at all. Instead, they were remembered as accurately as the angry white faces. In fact, when the task was especially mentally demanding—with the targets' faces being flashed very briefly alongside a distracting piece of art—the outgroup homogeneity effect was completely reversed, and angry black faces were more memorable than any of the whites.

These findings fit with the view that our brains allocate cognitive resources functionally—the mind frees up space to pay special attention to other people who might be especially pertinent to our survival or reproductive success. The reversal of outgroup homogeneity does not mean that we become less prejudiced when we are feeling threatened, only that threat makes us process information in ways that best serve our interests. Indeed, other research conducted by our team and by several different groups of researchers suggests that these same self-interested processes often boost stereotyping and prejudice.

Functional Projection

One of Sigmund Freud's many fascinating ideas was the concept of mental "defense mechanisms." Freud thought of defense mechanisms as tools we use to protect ourselves from anxiety. If some unpleasant memory is upsetting to you, for example, you can defend your ego by repressing it from consciousness or by denying that the unpleasant event ever happened. One of the more interesting defense mechanisms is projection—the inclination to attribute one's own unacceptable impulses to somebody else. Consider the oft-heard claim, "I'm not prejudiced at all"—but those damned Muslims/Christians/Jews/Protestants, they sure are!

Although Freud thought of projection as a neurotic means of protecting the self from anxiety, my colleagues and I thought that it might take other forms. In a series of studies in our labs, we examined a process we called "functional projection," or the tendency to project feelings onto others in ways that best serve our own adaptive goals. Unlike Freudian projection, in which I see my own undesirable feelings in others, functional projection may lead me to see other people as having very different feelings from my own. So, for example, if I am feeling fear, it would be functional to perceive that other people are feeling anger, especially if those other people might do me harm. We would expect this projection process to be biased in ways most likely to further my survival and reproduction. And it ought to be a very directed process—I ought to perceive the functionally important emotion only in people likely to pose specific threats or opportunities. It might also play a significant role in prejudice.

My colleagues and I wanted to know just how functional such projection could be, and we wanted to see what got projected and who it got projected onto. To find out, we showed white students facial photographs of black and white men and women. If you were a subject, you would have first heard an elaborate cover story: that you would see pictures of other people who had been photographed after having been asked to first think about a time in their life that had caused a strong emotional reaction and then to put on a neutral facial expression to hide the emotion we had evoked. Your job would be to detect the hidden emotion in the photographs. This might seem difficult, you would be told, but people can often subconsciously notice subtle micro-expressions on other people's faces. We would have also told you that you would do best if you based your judgments on your immediate gut reactions.

In reality, the task had nothing to do with the ability to detect emotional micro-expressions, because all the photos were handpicked to be emotionally neutral. We were really looking to see how a person's emotional states influenced what he or she projected onto the neutral pictures. To manipulate those emotional states, we showed each subject a movie clip, with the instructions to try to imagine what the main character was feeling. One clip was a scene from *The Silence of the Lambs* in which a white male serial killer stalks a white female FBI agent through a lightless basement. The scene ends with the killer, wearing night-vision goggles, reaching out to touch the woman, who is completely unable to see his hand approaching. A second clip, from *Things to Do in Denver When You're Dead*, depicted a highly attractive woman on a first date with a highly attractive man. The third, control clip, from the film *Koyaanisqatsi*, consisted of time-lapsed scenes of people going up and down on an escalator, working on an assembly line, or taking part in other activities typical of modern life.

The emotions that our subjects "saw" in the photographs depended partly on what they were feeling, but also partly on the people in the photographs they were rating. Men who were shown the romantic movie clip tended to overperceive sexual receptivity, and to do this only for women who were physically attractive. Conversely, women who saw the same clip did not project sexual emotions onto the people in the photos, even when those people were good-looking men.

Functional projection processes also led to a very specific type of racial stereotyping. The subjects who saw the frightening clip did not project their own fear onto the photographs, as a Freudian might have predicted; instead, they projected anger, and only onto black men, whom those white students associated with physical threat (which we knew from other surveys). And in another study in the same series, participants judged Arabs, and again projected not fear but anger onto the Arabs. We also measured people's implicit attitudes toward Arabs by measuring the speed with which they learned to associate positive and negative words with Arabs. In this study, only participants who had implicitly negative attitudes toward Arabs did the projecting, and they saw anger in Arab women as well as men. Instructively, this research was conducted at a time when news reports were full of stories of Arab suicide bombers, many of whom were women.

Our inclinations to use stereotypes can also be magnified not only by our internal states but also by environmental cues that amplify fear. Mark Schaller, Justin Park, and Annette Mueller, members of our team at the University of British Columbia (UBC), demonstrated how this works in a study of stereotyping in the dark. Nighttime is universally associated with evil, threat, and danger, which is not surprising: The threat of ambush is greater when you cannot see, so the evolutionary rewards of being especially wary at night can be substantial. Schaller and his colleagues wanted to see if simple darkness could trigger our use of self-protective stereotypes.

The study had two parts. First, Schaller and his colleagues measured their subjects' beliefs about how dangerous the world is, using a scale developed by Bob Altemeyer, a researcher who studies the links between personality and prejudice. It is pretty straightforward: Subjects describe the extent to which they agree with statements such as "Every day, as our society becomes more lawless and bestial, a person's chances of being robbed, assaulted, and even murdered go up and up." Schaller's team later tested their subjects to see what kinds of emotions they would project onto neutral photographs of black or white men. They team found that Canadian students who were chronically worried about a dangerous world were more likely to see the black men in the photographs as threatening, but only when the subjects were looking at the photographs in a darkened room. Darkness did not make the subjects more likely to express general prejudice; they did not agree more with other stereotypes, such as the purported ignorance or poverty of black men. Heightened fear simply lowered their thresholds for feeling threatened, especially by strange men from other racial groups.

These findings link up with some fascinating recent evidence on the neuroscience of prejudice. Using fMRI, a technique that directly measures ongoing brain activity, Elizabeth Phelps, Mahzarin Banaji, and their colleagues recorded white students' brain activity while they looked at photographs of black men. The researchers found that students with implicitly negative attitudes toward blacks showed high levels of activity in the amygdala (an area of the brain associated with emotional evaluation), but only when viewing photos of strange black men, not of famous ones such as Will Smith or Denzel Washington.

When Foreign Equals Disgusting

Mark Schaller is an interesting fellow. He spent his formative years in places like India and Africa with his father, George, a field biologist famous for his work with lions, gorillas, snow leopards, pandas, and other disappearing mammalian species now only found at the edges of human civilization. As a consequence, Mark is pretty comfortable with things that might seem extremely strange to those less adventurous folks, like me, who think of Italy or the Netherlands as an exotic destination. Mark's partner Quincy Young also had well-traveled parents, and she had spent her early childhood in Ethiopia. When Mark and Quincy's daughter was less than a year old, they not only hauled her into the Peruvian jungle on a backpacking trip but also took her to live for several months in Sri Lanka. When I expressed horror at their plan to take an infant into a third-world country, exposing her to God knows how many rare and exotic diseases, Mark just laughed.

Mark once had Paul Rozin—a researcher well-known for his studies of food preferences and disgust—over for dinner. As they were eating a salad from Mark's garden, they discovered a large and colorful beetle on one of their plates. Paul, who coined the term "omnivore's dilemma" to describe the tension between needing to find new foods and being disgusted by them, jokingly challenged Mark to eat it. So Mark picked up the beetle, put it in his mouth, and gulped it down.

This is all to say that Mark is not the kind of person who is easily put off by things that might disgust others. Most other people, however, are precisely that kind of person. In fact, Mark and his colleagues Jay Faulkner, Justin Park, and Lesley Duncan have observed that most people's minds equate "foreign" with "disgusting," and often associate foreigners with disease vectors, such as rats and lice. Ancient Romans equated foreigners with detritus and scum, for example, and during the recent genocide in Rwanda, Hutus referred to Tutsis as cockroaches. When I was growing up in New York, I often heard people justify their prejudices against Puerto Ricans and blacks with accusations that "those people" actually preferred living in unclean conditions. Ironically, I often heard those accusations from people whose own great-grandparents had been demonized in the same way.

This sort of bigotry may once have been functional. A strange person would have been more likely than someone from the local village to carry a disease against which our ancestors had no defense. So avoiding strangers might have helped avoid the latest version of smallpox, plague, or swine flu. If you have read Jared Diamond's *Guns, Germs, and Steel*, you know that many more Native Americans were killed by European diseases than by European guns. Of course, all evolved inclinations involve trade-offs. Our ancestors exchanged goods with members of other groups and often found mates from outside their own villages. Because total isolation means forgoing opportunities as well as dangers, Schaller and his colleagues reckoned that disease-avoidance mechanisms should be flexible. Glaring symptoms of illness or news of an epidemic would probably be enough reason to shun a stranger, as would one's own personal vulnerability to disease (for example, pregnant women who are carrying developing fetuses pay an especially high cost if they catch a disease). Otherwise, though, Schaller and his colleagues reckoned that it would not pay to be a xenophobe.

People vary in the extent to which they feel vulnerable to disease. Some people, like my friend Schaller, feel their immune system will take care of them and do not worry about someone else's slurping from their drinking glass or a stranger shaking their hand. Schaller and his colleagues developed a psychological scale to measure that sense of vulnerability. I scored high on the test, and when I took it, I was reminded of Schaller snickering at me during a dinner at a Chinese restaurant when I made a grimacing request that people not dig into the communal plates with their saliva-sodden chopsticks.

To examine the links between felt vulnerability to disease and prejudice, Schaller and his students asked Canadian students their attitudes about allowing an immigrant group into Canada. Sometimes the immigrants were described as coming from places UBC students might be likely to regard as unfamiliar and foreign, such as East Africa, Sri Lanka, or Peru; other times the students, who are mostly of European and Asian descent, were told that the immigrants were from more familiar countries in Europe or Asia. Repeatedly Schaller found that students who viewed themselves as chronically susceptible to disease were more xenophobic toward the unfamiliar groups than they were toward Europeans or Asians from more familiar countries.

Schaller's team also conducted a pair of experiments to see if priming subjects to worry about disease could affect their opinions on immigrants. Some subjects viewed pictures related to disease (such as images of bacteria on human hair and dirty kitchen sponges). The control group saw photos evocative of accidents in everyday life (such as having a radio fall into a bathtub). Then the participants were asked their opinions about immigrants from either a familiar country (such as Scotland) or an unfamiliar one (such as Nigeria). Those who had been primed to think about disease were more xenophobic toward people from unfamiliar countries.

Another study made a fascinating extension of these results by looking at the behavior of pregnant women. During the first trimester of pregnancy, a fetus is especially at risk if the mother contracts a disease. Women have a number of biological and psychological mechanisms designed to reduce these risks; for example, pregnant women become highly selective about their diet, avoiding novel foods and foods that are likely to carry bacteria, such as meat and fish. Women in the first trimester are especially prone to nausea and disgust. It is a burden, but a functional one: Women who suffer through more of these symptoms are less likely to have spontaneous abortions, and they tend to have healthier babies.

Based on what we have seen so far in this chapter, you might expect pregnant women to hold dimmer views of strangers. Carlos Navarrete, Dan Fessler, and Serena Eng decided to find out if that is true. To do so, they conducted a study in which they asked American women to evaluate two essays, one written by an American and expressing strongly pro-American opinions, the other written by a foreigner critical of the United States and its people. Pregnant women were four times more favorable than nonpregnant women toward the American. Furthermore, this relative antipathy toward the foreigner was substantially higher during the first trimester. Indeed, the incidence of xenophobic attitudes dropped over the course of pregnancy, just as incidence of the nausea did.

Race and Politics

All this research suggests that an evolutionary perspective can help us better understand stereotyping and prejudice. The more we understand these processes and what might magnify them, the better equipped we will be to combat them.

Combating prejudice has long been one of psychology's most noble goals. Unfortunately, the theories psychologists relied on were often incomplete and were sometimes just dead wrong. For example, before psychologists began to understand that the mind uses different rules for handling different kinds of information, we tended to talk about all forms of prejudice as being more or less the same thing—"negative feelings" about the members of a group. As my colleagues Steve Neuberg and Cathy Cottrell have pointed out, though, negative feelings come in more than one flavor. The same fellow might hold prejudices against black men because he fears a physical threat from the gang members on the corner, prejudices against homosexuals because he feels disgust at the thought of two men dancing and kissing, and prejudices against Asians because he worries they will out-compete him for money and jobs. As Schaller's research demonstrates, some prejudices might be motivated by a fear of disease and others by a fear of physical danger, so that prejudice against a black woman from Nigeria might have different motivations than prejudice against a black man from east Los Angeles. Only in some cases are two groups subject to the same prejudice, and sometimes in unexpected ways. For example, Cottrell and Neuberg found that college men and women had surprisingly similar negative feelings about fundamentalist Christians and radical feminists—in this case, students perceived both groups as threats to personal freedom.

If you want to reduce prejudice, then, an intervention designed to reduce the wrong kind could fail or even backfire—increasing the very prejudices you hoped to reduce. So this stuff really matters. Steve Neuberg and Mark Schaller tried to spell all this out in a paper they submitted to *American Psychologist*, the official journal of the American Psychological Association (APA). But the paper was rejected, and the reviews revealed all the old political antipathies toward evolutionary psychology. Despite the expressed intention of the APA to combat stereotyping, prejudice, and discrimination, an evolutionary approach to the matter was judged to be "insufficiently sensitive."

To be fair, *American Psychologist* has high standards, and most papers submitted there are rejected, so perhaps Neuberg and Schaller's paper was turned down because of other problems. But I did think that Neuberg and Schaller's paper proposing an evolutionary model of prejudice was much more important in its implications than many of the more "sensitive" papers published around that time (which frequently rehashed the claim that psychologists are subtly racist or sexist even when they are trying not to be).

The Fallacy of Biology's "Right Wing"

The controversy that dogs evolution-inspired papers like Neuberg and Schaller's flows directly from the controversy that dogged sociobiology in the 1970s. Ullica Segerstråle, a scientific historian who was working on her doctorate at Harvard when the firestorm began, wrote *Defenders of the Truth*, a detailed account of the earlier period. When Wilson's book *Sociobiology* came out, a group of Marxist intellectuals—including Richard Lewontin, a famous Harvard biologist, and his colleague Stephen Jay Gould, a paleontologist who became a widely read science writer—mounted a heated attack on the idea that human behavior could be understood in evolutionary terms. The opponents argued that sociobiology was an attempt by white male elitists to justify the status quo. Indeed, Gould argued that sociobiology was linked to Nazism, anti-Semitism, sexism, and other forms of societal evil. After Gould died, the torch passed to several zealous followers, including Hilary and Steven Rose, who claim that evolutionary psychology "is transparently part of a right-wing libertarian attack on collectivity, above all the welfare state."

A cornerstone assumption of this critique is the claim that evolutionary accounts of behavior include an intrinsic assumption of genetic determinism, which in turn precludes societal change. If this were true, then it would hardly be appropriate for *American Psychologist* to publish an article on evolutionary psychology and racial prejudice. It would be worse than insensitive; it would be aiding and abetting a right-wing plot to convince people that racism is in their genes and therefore impossible to change. The problem with the critique, however, is that neither the premises nor the conclusions have a shred of truth.

For one thing, even if evolutionary theorists were in fact a pack of right-wingers, or even a pack of mildly conservative people who like the status quo, that should have no bearing on how one evaluates their findings. Science proceeds by challenging our assumptions, whatever they may be, and then submitting those challenging alternatives to empirical tests. Good reasons for publishing or not publishing a research finding ought to include whether it is interesting and whether the data supporting it are rigorous—not whether the scientist based his or her hypothesis on the proper political ideas. But I am sort of wasting my breath here, because as it turns out, evolutionary psychologists are, like most academics, a left-leaning group. I have been to several meetings of the Human Behavior and Evolution Society, and the people I see there are a lot of young women and men who look as if they belong to the Sierra Club and hang out in coffee shops that sell fair-traded organic coffee. Most of the old white men there tend to sport the "I was a hippie in the '60s" look.

In fact, when Josh Tybur, Geoffrey Miller, and Steve Gangestad surveyed a sample of 168 doctoral students in psychology on political issues and their votes in the 2004 presidential election, they found that evolutionary psychology students were substantially more liberal than the American populace, and even more liberal than other academics. At the time of Tybur's survey, 30 percent of the American population identified as Republican; none of the evolutionary psychologists did. In this, they were similar to other psychologists, only slightly more liberal (12 out of 137 non–evolutionary psychologists, but none of the evolutionary psychologists, had voted for George W. Bush). When questioned on specific beliefs, evolutionary psychologists also expressed significantly more liberal attitudes than the U.S. population, and again they were like other psychologists, only slightly more liberal. The only substantial difference between evolutionary psychologists and the other group was that evolutionary psychologists were significantly more positive about the use of the scientific method.

As for claims of genetic determinism, these are also simply false. Indeed, research generated by evolutionary psychologists has suggested something very different: Evolved psychological mechanisms were designed by natural selection to respond to *variations* in the environment. Hence, evolutionary psychology is inherently concerned with discovering the varying environmental cues that turn adaptive mechanisms on and off. All the findings discussed in this chapter reflect this. I have talked about findings showing that certain kinds of prejudice are triggered by fear, others are triggered by concern with disease, and still others by economic threats. And the same is true for all the other chapters in this book: We have been talking a lot, and will talk more in later chapters, about how adaptive psychological mechanisms respond to variations in threats and opportunities in the environment. That just does not qualify as genetic determinism. (If you want more evidence, read John Alcock's *Triumph of Sociobiology* or Steven Pinker's *Blank Slate.*)

At the same time, seeking to understand the mechanisms that cause prejudice is not the same as justifying them. Those who claim otherwise are falling for something called the naturalistic fallacy, which confuses discussions of what is "natural" with discussions of what is "good." It is easy to see why one might make this mistake if one does not think too deeply about it. Indeed, the word "natural" often does have a positive connotation, as in "natural foods." However, the positive connotation does not quite hold up if you look at the natural world. Here are some things that are quite natural: tuberculosis, cancer, AIDS, malaria-bearing mosquitoes, leeches, tapeworms, infanticide by male lions, earthquakes, and tsunamis. None of these natural things would typically be regarded as intrinsically morally superior to "unnatural" things like iPods, impressionist paintings, hybrid cars, or well-manicured English gardens.

The same holds for human behaviors—natural does not mean good. Here are some things that evolutionary psychologists claim are natural: men's tendency to commit more homicides than women, women's inclination to be more unfaithful while they are ovulating, and, as I just discussed, prejudice against members of completely innocent people from Harlem or Sri Lanka triggered by fear or concern over disease. If there is a moral case for these behaviors, I have never heard it coming from an evolutionary psychologist. But they would claim that these behaviors are the result of natural processes, just like maternal love, the inclination to share, and a concern for justice. Thinking about the ways in which we are naturally nice demonstrates that evolutionary analyses do not lead to an especially negative view of human nature, just to a neutral view.

If You Want to Fight Serpents, You Have to Kick Over Rocks

If you want to live in a nicer world, you need good, unbiased science to tell you about the actual wellsprings of human behavior. You do not need a viewpoint that sounds comforting but is wrong, because that could lead you to create ineffective interventions. The question is not what sounds good to us but what actually causes humans to do the things they do.

As it turns out, some of the findings coming out of evolutionary psychology labs suggest grounds for optimism. Babies are not born knowing whom to regard as their enemies; they have to learn whom to associate with fear. In our research on angry faces, for example, we found that American students were not especially likely to associate Asian faces with threat. In fact, white people continue to homogenize Asian men when they are expressing an angry expression.

As Rob Kurzban, Leda Cosmides, and John Tooby have noted, our ancestors were very unlikely ever to come in contact with members of other races. Instead, they came into conflict with people living in the village just down the river—people who probably looked a lot like they did. In a very interesting study, Kurzban and his colleagues showed students a highly competitive basketball game in which opposing teams wore different-colored shirts. The observers never confused men and women with one another, but they did mix up black and white players if they happened to be on the same team. So it looks as if race is fairly easy to erase from people's minds. We saw that dynamic at work during the 2008 presidential election, when Republicans and Democrats alike were more likely to think of Joe Biden as a member of Barack Obama's tribe than of John McCain's.

So the bottom line is this: Not only can an evolutionary perspective help us understand why humans are so universally inclined to feel prejudice toward members of other groups, but it can also help us understand the factors that make the strength of those inclinations go up and go down. If you happen to believe that intergroup prejudice is a bad thing and hope to discover how to reduce it, as most psychologists do, then you should not close your mind to the actual underlying mechanisms. One could certainly ask whether the academics who wrongly accused a group of left-leaning evolutionary theorists of being similar to Nazis were especially sensitive, but don't get me started.

Instead, let me close this discussion on a more optimistic note. In thinking about evolution and prejudice, my colleague Mark Schaller likes to toss around a pair of quotations from Barbara Kingsolver

(who was a graduate student in evolutionary biology before she became a novelist):

> We humans have to grant the presence of some past adaptations,
> even in their unforgivable extremes, if only to admit they are per-
> manent rocks in the stream we're obliged to navigate.
>
> A thousand anachronisms dance down the strands of our DNA
> from a hidebound tribal past. . . . If we resent being bound by these
> ropes, the best hope is to seize them out like snakes, by the throat,
> look them in the eye and own up to their venom.

Shifting Our Gaze Out of the Gutter

So far, we have been hanging out in the gutter, talking about sex, aggression, and prejudice. Now we will shift our gaze higher, considering how the findings of evolutionary psychology reflect on two
profoundly important philosophical questions about human nature
and how the mind works. But if you prefer pondering the many
puzzling aspects of sex and aggression to an abstract philosophical
treatise, don't turn that dial yet. We will consider the question of
human nature and the mind in light of some very interesting findings on male-female relationships—in this case, romantic affairs between younger women and older men. For decades, social scientists
thought the phenomenon was a simple and obvious outgrowth of
American culture. But the real reason turned out to be neither simple
nor obvious. In fact, uncovering the root cause of men's and women's
age preferences has involved something of a scientific mystery story,
complete with a search for clues that took my colleagues and I around
the world, and with bad police who wanted to halt the investigation.

Chapter 5

THE MIND AS A COLORING BOOK

Bozeman, Montana, 1977. I'm a brand new assistant professor at Montana State University, land of *Zen and the Art of Motorcycle Maintenance*, where the snow-capped Bridger Mountains are visible from campus. I've just given a lecture to a faculty group about my research on human mate preferences, and it's the question-and-answer period. An anthropology professor in the audience is addressing me with a rather stern tone in her voice. Like many academics, she has acquired the peculiar custom of stating her "questions" in the form of lectures. These lectures-disguised-as-questions are usually delivered in a haughty and didactic tone, meant to set the speaker straight on some topic about which the questioner fancies himself or herself more expert.

In my talk, I had made a generalization about human mating strategies, which triggered a minisermon about the range of cross-cultural variations in human mating. The anthropologist seems quite adamant in her conviction that, as a psychologist who conducts laboratory experiments with a narrow sample of American college students, I have absolutely no justification for making any sweeping generalizations about humans as a species. When one looks across cultures, I am admonished, one finds an unlimited variety of male-female relationships. In other words, she is explaining why and how anthropological research has proven that the mind is a blank slate. But has it really? Some research I conducted on dirty old men's preferences for younger women has made me increasingly dubious of this claim.

Middle-Aged Gent Seeking College Cheerleader

A few years after I left Montana State, I gave a lecture about interpersonal attraction to a singles' group in Phoenix, Arizona, and one of the older women in the group asked me why it was that men in her age group all seemed to be prowling around for "young chicks." The other older women all chimed in with their agreement, and as evidence, they handed me a pile of singles' newspapers. Many of the ads were written by men who listed their own age as being in the forties or fifties but who were seeking relationships with much younger women.

I lugged the newspapers home and showed them to my longtime friend and colleague Rich Keefe. Rich had been in graduate school with me at ASU, where we had studied how to apply behavioral learning principles to clinical psychology. Like me, Rich had come to believe that psychology needed to be updated by evolutionary thinking. We began to look at the ads to see if the women's complaints held up, and if so, to ask what evolution could tell us about it.

Several social scientists had previously analyzed data from singles' advertisements and had noted that women, on average, sought men a few years older than themselves and that men sought women a few years younger than themselves. This was one of the few known exceptions to a social-psychological law called the similarity-attraction principle—the general tendency for people to desire friends and romantic partners who are carbon copies of themselves. For example, liberal, Jewish, nonsmoking mountain-bikers are usually seeking a partner with the same interests and attributes rather than hoping to expand their horizons by dating conservative, Baptist, chain-smoking

Harley riders. Earlier researchers expected that the same rule would hold for age preferences, that older men and women should prefer older partners.

When they discovered that age preferences violated the similarity rule, the researchers blamed American cultural norms for the discrepancy. For instance, sociologist Harriet Presser suggested that there was a "norm" that "a husband should be, or at least appear to be, mentally and physically superior to his wife. Not only should he be taller than she (for the appearance of superiority) but also older (which gives him the advantage of more time to become better educated and more experienced)." Along similar lines, psychologist Leticia Anne Peplau and sociologist Steven Gordon put it this way: "American culture encourages sex-linked asymmetries in the characteristics of dating and marriages" in which, for example, "women are traditionally taught to seek a man who is taller, older . . . more occupationally successful."

Around the same period, Julie Connelly published an article in *Fortune* magazine in which she used the term "trophy wives" to refer to attractive younger women who were the second wives of older, powerful American executives. When social scientists were called on to explain the phenomenon, they ascribed it to something about modern American culture. One sociologist, for example, attributed the trophy wife syndrome to cultural images in the media, which depicted the ideal man as a successful businessman in his late forties or fifties, and the ideal woman as an ingenue in her twenties or early thirties.

Reexamining the Evidence

Keefe and I seriously doubted that the older man–younger woman phenomenon was really a product of American cultural norms or modern media images. We thought instead that it could be explained in light of a couple of universal biological differences between women and men. First, women undergo menopause, a complete cessation of fertility, during their forties. Men do not. On the other hand, women are highly fertile during their twenties, and the features men find attractive in women, such as rounded hips, full breasts, and lustrous hair, are indirect cues to that fertility. A strong innate bias for those fertility cues would easily account for the older men's preference for younger women. On the other side, we suspected that women are seeking men who could contribute indirectly to their children by providing food, protection, and other resources. To the extent that men continue to accumulate resources and social status with age, women would be expected to prefer older men.

Our theory did more than just reexplain the existing findings; it had new and testable implications. If we were right, and men were seeking fertility, then the preference for younger women would be very strong only in older men, not in very young men (because for men in their late teens and twenties, their age mates are highly fertile). But to test our ideas, we had to look at age preferences in a different way. Earlier researchers had simply clumped together advertisers of all ages and reported an average age discrepancy for each sex. We instead separated the advertisers into age categories, and this revealed a pattern more complex than men wanting slightly younger women and women wanting slightly older men. Indeed, the pattern Keefe and I discovered fundamentally contradicted the standard social science explanations of why men and women act like they do.

Women's preferences were not the problem: Keefe and I found that women were acting just as earlier researchers had described. Women were looking for somewhat older men, and this general pattern persisted throughout their lives. We were actually surprised to find that the preference for slightly older men even persisted among women in their sixties, when there are a lot fewer older men to choose from.

The men's preferences, however, shifted dramatically according to the age of the guys. The youngest men, despite the supposed societal expectation that they should look for younger women to dominate, were instead interested in a range of women. A typical guy of twenty-five was interested in women as young as twenty and as old as thirty. In a later study, we found that teenage boys were most attracted to women slightly older than themselves—college-age women. Teenage boys expressed this preference even though they realized the older women were unlikely to reciprocate their interest. But as men aged, this preference for partners their own age progressively shifted to an interest in women younger and younger than themselves. A typical forty-five-year-old guy wanted nothing to do with women his own age; instead, his preferences ran to women five to fifteen years younger. And men of fifty-five were even more extreme in their desire for younger women. In this sample from the late 1980s, men who came of age listening to Elvis Presley were eying the girls on their way to U2 concerts, while the women of the same rockabilly generation were hoping to woo a codger from the Frank Sinatra era.

As one critic suggested, the singles' ads might represent nothing more than fantasy. Who cares what people say, the argument went; whom do they actually pair up with? Maybe the president of the Acme Widget Corporation can attract a younger woman, but the average Joe down on the receiving dock can only dream. So Keefe and I collected a random sample of marriages from Phoenix, the city from which I had gotten the first samples of singles' ads. In a pattern exactly matched to the singles' ads, age discrepancies between men and the women they married got bigger with the man's age. Younger men married women near their own age, and a reasonable number of young men married women slightly older than themselves. Older men married women increasingly younger than themselves—just like the rich and powerful CEOs of Fortune 500 corporations.

Stated simply, then, our findings suggested that people's apparent age preferences are not ultimately about age at all. Because women contribute their own bodily resources to the offspring, men are seeking

cues linked to fertility and health. Because men contribute indirect resources to the offspring, women are seeking cues linked to the ability
to acquire those resources. Men's resource acquisition and women's fertility are correlated with age, but age itself is not the driving force.

To someone with any background in evolutionary biology at all,
our explanation might sound self-evident. But at the time, most social scientists did not think much about a link between human
courtship and reproduction. And when we started talking about our
findings in these evolutionary terms, we were often met with derisive
sneers and claims that our account was obviously wrong, along with
arguments that our findings could be explained in terms of the
"norms of American culture."

Of course, there are norms in American society indicating that it is
less typical and appropriate for a woman to pair off with a younger
man than with an older man. But even if a behavior is consistent with
a norm, that is no proof that the norm caused it. Sometimes a norm
is prescriptive (Thou shalt not marry someone under the age of con-
sent, for example), sometimes a norm is simply descriptive (men fantasize about sex more frequently than women do, but not necessarily
because they were told they had to do so).

It is easy to generate alternative explanations for almost any phenomenon. The trick is finding evidence that reflects on those opinions. In the remainder of this chapter, I will describe our search for
further evidence that could distinguish between the normative and
evolutionary explanations for sex differences in age preferences, as
well as a few surprises that we discovered along the way.

Searching for Dirty Old Men Across Times and Cultures

The data we had found so far was interesting but insufficient to dis-
miss the possibility that all modern Americans, teenage boys as well
as aging CEOs, are influenced by exposure to similar cultural norms

and that the mass media were the vehicles for disseminating those norms. One way to address the "modern media" hypothesis was to examine marriage data from before the advent of television. To that end, we examined data from Phoenix marriages for the 1920s. Television was invented during the 1920s but was not commercially available until the 1940s, and fewer than 1 percent of homes had a television before 1950. The first TV station in Arizona was not even licensed until 1949. But Keefe and I found that marriage patterns were the same in the good old days, long before Ricky Ricardo fell in love with Lucy. Indeed, the pattern from the tubeless '20s matched perfectly with the one we had found in all the data from the 1980s, including the CEO marriages.

It was still possible, though, that the phenomenon might be linked to a different element of American culture, something that predated the modern media age—materialism perhaps, or individualism, or any of a host of other features of American society. To determine whether some common American cultural element underlies this pattern of behavior, we began to examine similar data from other societies. Our colleague Guus Van Heck from the University of Tilburg sent us data from singles' ads in newspapers in the Netherlands, and Ute Hoffman and Kirstin Schaefer of the University of Bielefeld sent us the numbers from singles' ads in Germany. In both cases, the data revealed the same sex-differentiated pattern we had seen in the United States. And the pattern was again not limited to Euro singles' fantasies or to the modern media age; Sarynina Nieuweboer uncovered the same pattern in actual marriages in Amsterdam from the seventeenth, eighteenth, and nineteenth centuries.

But it was still possible to argue that the Netherlands, Germany, and the United States all have European-based cultures and might therefore be subject to many of the same normative pressures. We were fairly confident that this was not a limited cultural pattern, but we needed to prove it. So we looked next at several non-European societies. We began with marital advertisements from Indian news-papers, which my colleague Steve West first brought to our attention. The Indian marital ads indeed painted a picture of a very different society, with very different cultural norms about marriage. The ads were commonly placed by family members acting on behalf of the unmarried individual, and the ads specified characteristics related to caste, subcaste, and subdenomination of the Hindu or Muslim faith. They also typically requested horoscope information. For example, one ad placed in Bombay's *Times of India* stated:

> Wanted: a non-Bharadwaj smart good-looking preferably employed Kerala Iyer girl below 25 for a Kerala Iyer boy 29. Chemical engi-neer. Contact with horoscope.

Despite the abundant cultural differences apparent in these Indian marital ads, we nevertheless found the same pattern of sex differences we had found in North American and European samples. When an Indian woman's relatives advertised to find her a husband, they asked for a slightly older man regardless of her age. As Indian men aged, their relatives sought women who, relative to the men, were progres-sively younger and younger.

But wait. India is not a European society, but it was for a time under the rule of Britain. Although the marital ads demonstrated that British rule had not eradicated many of the central features of Indian culture—such as the caste system and the emphasis on the Hindu Zodiac—a skeptic could still argue that the preference for young women to marry older men was a result of British rule. That possi-bility, however, seems increasingly unlikely in the face of evidence from a range of cultures; similar age differences in mating patterns have now been reported among Brazilians, African pastoralists, and Pacific Islanders. For example, Nenita Estrera and her students sent us data on marriage ages recorded from 1913 to 1939 in a remote

Philippine fishing village called Poro. Poro marriages fit perfectly with the modern American pattern, and with the Dutch, German, and Indian data. Indeed, older men on Poro married even younger women than their counterparts in America, a detail that poses a particular problem for those who would attribute the phenomenon to American media images. It became clear that the phenomenon of older men marrying younger women is not limited to American culture or to the modern era.

As a younger fellow, incidentally, I had high hopes of not growing into a dirty old man. As it turns out, I failed miserably, but in a way that made me one more data point in this survey. My first wife, whom I married in my early twenties, was born the same year as I was—a fellow member of the generation of college students who listened to Jimi Hendrix and Frank Zappa at 33⅓ rpm. My second wife, whom I married in my late thirties, was a decade younger, educated with ZZ Top and Steely Dan playing in the background. My third (and I hope my last) wife, whom I married in my mid-fifties, is still another decade younger, from the generation that gave the world Vanilla Ice and the Smashing Pumpkins.

A Startling Exception That Proves the Case

I began this book describing how a random work-avoiding visit to the school bookstore led to a revolutionary change in my worldview—when I stumbled across Jane Lancaster's evolutionary account of primate social behavior and the emergence of human culture. Many years later, after I had published my findings on the universal attraction between older men and younger women, I was rummaging through the anthropology section of a used bookstore. I chanced to pick up a dusty old ethnography, titled *The Tiwi of North Australia*, by the anthropologists C. W. M. Hart and Arnold R. Pillig. I am fairly certain that the other bookstore customers heard me gasp as I began reading about this group of aboriginal Australians. The Tiwi's customs seemed to call for another worldview disruption, for they seemed to challenge my view that the attraction to younger women was universal, and to support the view that the mind was a completely blank slate.

As the authors of the Tiwi account noted:

> According to a nearly complete genealogical census carried out in 1928–1929, nearly every man in the tribe in the age group from thirty-two to thirty-seven was married to an elderly widow. . . . But very few of them had a resident *young* wife.

A society in which men were more attracted to postmenopausal women than to young fertile women was shocking for more than one reason. It posed a problem not only from the perspective of evolutionary psychology, but also from the perspective of Biology 101. How could the members of such a society reproduce themselves?

I began to dig more carefully into the Tiwi ethnography. A closer examination revealed that Tiwi society was distinct in a number of interesting ways. But although young Tiwi men do in fact marry older women, they are, like men in other societies, attracted to young women as sexual partners. Indeed, Tiwi men are obsessed with younger women. A big problem in Tiwi society was keeping unmarried young men away from young women, and there were stern rules designed to accomplish this end. If a young guy was caught messing around with a young woman, he could be gored with a hunting spear, or he could be expelled from the group (which at that time in northern Australia may have been tantamount to a death sentence).

Why all the concern about separating the young men and women? The answer lies in two other features of Tiwi society, and those features also resolve the more troubling biological question of how the Tiwi reproduce themselves if all the young men are married to elderly women.

First, let us ask, Where are the fertile young Tiwi maidens while all the young men are marrying elderly women? Are they waiting around till they reach forty-five or fifty years of age for an opportunity to marry a twenty-five-year-old man, who will take no bride before she fully matures? No. The young maidens are all already married. In fact, all the females in Tiwiland marry quite young—at birth, in fact! Whom do they marry? The old powerful patriarchs—who rule with a firm hand and use their extensive power to monopolize every single one of the young women. When a Tiwi girl is born, she is immediately betrothed. Her father chooses her husband, and according to anthropologists Hart and Pillig, the old man considers her "an asset . . . to be invested in his own welfare." What makes the daughter such an asset is this: The society is polygynous, and older men mostly betroth their young daughters to other patriarchs, who are in a position to reciprocate when one of their wives has a daughter. So the older men exchange younger women with one another, and the young blokes, with no daughters to offer, are out of the game, completely excluded from obtaining young wives who could bear them daughters.

As in all societies, young Tiwi men are nevertheless attracted to young Tiwi women, and young women do sometimes fall for young men (who are probably much more physically attractive than their elderly husbands). This allows the occasional young man to steal some contact with a young woman. But the Tiwi patriarchs are ever alert for such incursions into their monopoly on fertile women, and they enforce severe punishments on any young Romeo caught running around with a fertile young Juliet. The standard punishment is this: The young man must stand in the center of the village and allow the elder cuckold to throw a spear at him. The young fellow can jump out of the way, but then the elder gets another shot, and another, until he scores a hit. Another troubling twist for the young man: If he keeps jumping completely out of the way, the other elders pick up their spears to help their fellow patriarch save face, and then the young man confronts a shower of spears coming at him. Ideally, the young fellow allows the wronged old man to hit him in the leg, sheds some blood, and the case is closed. But sometimes he does not jump in the right direction, and the spear pierces his upper body, so the young fellow can die from the wounds. (Even a leg wound was no laughing matter in a time and place with neither antiseptics nor antibiotics.)

The harsh patriarchal sanctions explain why younger men and women do not marry, but not why young men and older widows marry. Other features of the social system help solve the puzzle. To allow themselves to control all the young brides, the older men enforce a rule requiring that all Tiwi females (but not all males) be married. So as I mentioned, a girl is betrothed to an older man as soon as she is born. But there is another side to the female marriage rule: As soon as an older man dies, his widow or widows must remarry. The powerful older men, who frequently have numerous young wives, are not interested in marrying the older women. So who is an elderly widow going to marry? At this point, a younger man steps up. What's in it for him? By marrying a widow, the young man builds alliances with her relatives, and he gets the right to determine whom her younger daughters marry if they become widowed early (remember, all the young girls are also officially married to elderly guys, so when an old fella kicks off, he may leave some fertile younger wives as well as the older ones). Once a young man marries a widow, then, he is in the game. His status goes up, and he becomes eligible to acquire younger wives.

So instead of overturning the evolutionary life history model, the Tiwi pattern suggests a dynamic interaction between an evolved psychological mechanism (men's attraction toward women in the years of peak fertility) and local social ecology (a geriatric patriarchy that monopolizes younger women for themselves, in combination with a rule that all women must be married).

Blank Slates, Jukeboxes, and Coloring Books

The Tiwi case is just one example demonstrating that the human mind is hardly a blank slate when it comes to absorbing and constructing cultural practices. Although human societies vary in numerous ways, those variations are not infinite and are not random, and they do not typically violate general principles that apply to all animal species. Although there are a few intellectual holdouts, most social scientists today would agree that the mind is not well characterized as a completely blank slate. But despite the general agreement that John Locke's old metaphor is outdated, it has kept its appeal as a simple and memorable image. We ought to replace it with an image that is equally straightforward and understandable.

John Tooby and Leda Cosmides have suggested one interesting alternative: the mind as a jukebox. Compared with a blank slate, a jukebox is pleasingly interactive; what comes out of a jukebox is determined not solely by what is inside, nor by outside inputs alone (pressing F6 does not result in a tune unless there is a record inside corresponding to that button). The jukebox has one limitation, though: Many cultural norms are not straightforward, automatic consequences of pushing a particular set of preordained buttons. There is some flexibility, and the potential for unexpected combinations of different norms, as we saw in the case of the Tiwi.

I have suggested that we instead replace the blank slate with a view of the mind as a coloring book. A coloring book, like a jukebox, also paints an image of inner structure (the predrawn lines that suggest a giraffe rather than a zebra or a rocket ship) interacting with environmental inputs (the young artist wielding the crayons). But the coloring book metaphor has a few other advantages. For one, a coloring book leaves more room for flexible and potentially unpredicted outcomes—one child might choose to color his giraffe purple and green instead of tan and brown. At the same time, a coloring book paradoxically allows for more built-in constraints alongside its flexibility. More than the buttons on a jukebox, the predrawn outlines in a coloring book strongly solicit particular inputs from the environment (most children coloring a giraffe will be inspired to search for tan, brown, and yellow rather than purple, blue, and green). So, although a coloring book can, in one sense, be colored in an infinite number of ways, in another sense it is not really completely "passive" because the outlines strongly suggest particular palettes of inputs to be used on the different pages.

The coloring book metaphor does not pretend to be an actual representation of the human brain, but it does provides a straightforward contrast to the blank slate, conceptually extending this old and powerful metaphor to help us better visualize how mind and culture interact. Indeed, the coloring book actually incorporates the old metaphor, but it inspires us to think about the mind as having some built-in outlines as well as a great deal of blank space to be filled by environmental inputs.

An additional advantage of the coloring book metaphor is that, unlike a blank slate, a coloring book includes not just one page, but many. Just as there are different patterns of predrawn lines on different pages of a coloring book (tigers on one page, zebras on another, and giraffes on a third), so there are probably different constraints involved in solving the different developmental problems and opportunities involved in getting along, getting the bad guys, getting ahead, and getting the girl (or guy). In fact, as I will describe in the next chapter, another profound insight about the mind has arisen from the concordance of modern evolutionary psychology and neuroscience. Although it feels natural to think of our minds as unitary, it now appears that there are lots of different people running around inside our heads, alternative selves with very different, and sometimes incompatible, ideas about what to do next.

Chapter 6

SUBSELVES

In the summer of 1992, I traveled through Europe with my delightful son Dave, my close friend Rich Keefe, his fun-loving son Richie, and my good-humored second wife Melanie. Before we left, I had imagined a script for our travelogue that was *The Sound of Music* meets *Cinema Paradiso*: Ahh, intimate friends, drinking Belgian beer on the market square in Leuven, eating French baguettes by the Seine in Paris, biking beneath the snow-covered Swiss Alps, enjoying bubbly vino and fresh pasta on the quaint piazzas of Padova, Italy—it seemed hard to imagine a more joyous adventure.

But in fact, by the time we arrived in Paris at the end of the first week, it was closer to *National Lampoon's European Vacation* meets *Lord of the Flies*. The two teenage boys had teamed up against the adults, registering a steady stream of complaints about our various failings, including what they perceived as an authoritarian reluctance to buy them McDonald's hamburgers while forcing them to eat weirdly flavored French crap. And in such small portions! Between bouts of complaining, the lads were inclined to sleep till noon most days, so my wife took to leaving for museums on her own in the late morning (often in a huff). When we arrived in Paris, the rooms Rich had booked were gone. Hotels were packed, so we were forced to stand in a very long line in a very hot and crowded train station trying desperately to make alternative accommodations, speaking over the phone in broken French (understanding very little of the responses except "Non"), and trying to guard our luggage and wallets against the abundant local thieves (several of whom were arrested right before our eyes). We finally squeezed into a couple of overpriced, undersized, and underventilated rooms in a run-down hotel with a grumpy fellow at the desk, who acted as if it was a great burden to even hand us our room keys.

During our short stay in the ever-welcoming Eiffel city, my threshold for annoyance grew so low that I growled at a clerk in a bakery, "We should have left you bastards to the Germans!" This embarrassed my friend Rich, who got to be tainted by association when I stormed out of her shop. The bakerwoman's offense was that she had been disdainfully snotty, and although Rich pointed out that this was hardly grounds for resurrecting the Gestapo, her one slight was, in the context of an otherwise unpleasant string of experiences, enough to inspire me to wish ill fortune on her and all her fellow Gauls.

It was not just the snobby French who were getting on my nerves. Rich and I began to bicker over what should have been trivial matters—where to eat breakfast or what kind of bread to buy for lunch. During the third week, after we arrived in the otherwise charming northern Italian town of Padova, Rich and I spent one evening screaming at one another over a profoundly important issue: whose turn it was to wash the dishes in the (crowded and hot) apartment we were borrowing.

My son Dave was barely into his teens, and although he had previously been as easygoing a youngster as could be imagined, he chose this trip to try out for the role of stereotypical disaffected teenager. Dave spent most of one day walking half a block behind my wife and me, rolling his eyes skyward every time I spoke to him, communicating sheer disgust at any association with his grumpy old man. Demonstrating my mastery of adolescent psychology, I snapped at him, "It cost me $1,000 to buy your plane ticket; you could at least try to enjoy yourself!" His quick reply: "You shouldn't have wasted so much money. I'd be enjoying myself a lot more if I was back home playing basketball with my friends!"

After the nightmare European vacation, it seemed for a while as if my decades-long friendship with Rich might have been permanently damaged. And the trip aroused a level of annoyance between my second wife and me that had never existed before, a feeling that might well have contributed to our eventual divorce. But even though I viewed my son as one of the chief engineers on the Eurail Pass to hell, it did not put a dent in my feelings for Dave. In fact, after the dysfunctional larger group splintered, Dave and I spent the final week bicycling happily through northern Germany, glad to be spending time alone together (with the faint sounds of "Edelweiss" playing in the background on the movie sound track).

The very different consequences these unpleasant experiences had on my feelings toward my son, my best friend, my wife, and the anonymous French clerk are relevant to an important point made by evolutionary psychologists: The human brain does not use the same set of rules to make decisions about different people in our lives. Indeed, it may even be a mistake to talk about "the brain" as if it were one organ, somehow encompassing our tightly unified self. Instead, it makes more sense to imagine that each of us has a loose confederacy of subselves inside our head, each controlled by a different combination of neural hardware and software.

For most of the twentieth century, psychologists and other social scientists would have regarded what I just said as blasphemous—a willful violation of one of the cherished values of science. According to the much-revered criterion of parsimony, a scientific theory should strive to explain natural phenomena using as few assumptions as possible. Several highly influential twentieth-century theories promised to explain all human social life using a very small number of assumptions indeed—only one.

Simple Minds and Domain-General Theories

One of the admirably parsimonious theories in the field of psychology is something called the reinforcement-affect model. That theory assumes that people are motivated by one very simple goal—the desire to feel good—which inspires us to like people we associate with good feelings and to dislike people we associate with bad feelings. Love and hatred are the ends of this simple continuum, and whether we feel one way or the other about a person is happenstance, owing merely to what else was going on when we met him or her.

The reinforcement-affect principle had been invoked to explain a broad range of research findings: why we come to like other people who just happen to be around when we hear good news, why we like people who agree with our attitudes, and even why we are drawn toward people who are physically attractive. Good tidings, agreement, and beauty all make us feel good, and those good feelings rub off on anyone who happens to be around at the time.

The reinforcement-affect model is directly based on the simple laws of conditioning: Just as Ivan Pavlov's dogs came to associate a bell with food, so we condition good feelings to anyone who is around when something nice happens.

How do you get from simple conditioned reflexes like salivation in dogs to something seemingly more complicated, like love in humans? For this, the behaviorists made a distinction between primary and secondary drives. Primary drives include biological cravings like hunger and thirst, and secondary drives are desires for something that was previously associated with the satisfaction of a primary drive (if the friendly waitress at your favorite restaurant hails from Alabama, for example, you might develop a habit of drooling every time you hear a drawl). On this view, championed by B. F. Skinner, the desire for status or for love is not built in at the factory but is added on, and all the complexities of human life depend on two very simple forms of learning: classical and operant conditioning. During much of my graduate-school career, I was a fan of this elegantly simple viewpoint, which I found especially appealing because the rules of classical and operant conditioning applied in the same way to many kinds of learning across many species.

The reinforcement-affect model is a domain-general model—one that attempts to explain all behavior using a simple and parsimonious rule, in this case, "Do it if it feels good." But parsimony is not the only criterion for evaluating a scientific theory. A theory also has to do a good job of making distinctions when distinctions are necessary. For instance, several studies, including some I had done myself, had found that we like people more when we meet them under unpleasant circumstances, provided they are in the same boat and are not causing the unpleasant feelings. And sometimes the very same experience, such as seeing someone beautiful or handsome, can make one person feel good and make another person feel bad.

For example, Dan Montello, Sara Gutierres, Melanie Trost, and I did a study in which we asked people to look at a series of very attractive people and then report their mood. On the simple notion that it feels good to look at physically attractive people, that experience should have made our subjects feel good. For some of our participants, the last person they saw in the series was much less attractive than the others. Again, if attractiveness translates into positive feelings, the less attractive person should have brought people's mood down. In fact, sometimes it worked just like that: Looking at very attractive people in fact lifted people's moods, and seeing an average-looking person at the end of the series of good-lookers was a buzz-killer. But this was only true when the good-looking people were members of the opposite sex. On the other hand, seeing a stream of good-looking people of one's own sex can make the viewer feel somewhat crappy, and an average-looking person at the end of the series is a mood-lifter.

So to say that physical attractiveness is a "reinforcer" is not quite specific enough. This highlights a big problem with simple reinforcement models—they do not specify in advance which kinds of experiences are likely to be rewarding and which are likely to be punishing. Skinnerian behaviorists traditionally liked to define reward and punishment empirically: Something is a reward if an organism will work to get it and a punishment if an organism will work to avoid it. But this buys parsimony at the cost of circularity.

There is another interesting feature of our mood-attractiveness experiment, and it highlights a second critical problem with simple domain-general models of behavior. Even though an average-looking person at the end of a series of perfect 10s has opposite effects on men's and women's moods (depending on the sex of the people they see), both sexes made similarly negative judgments of those average-looking folks. That is, although an average-looking female makes women feel good when she breaks up a stream of fashion models, the good mood does not generalize to their ratings of the average-looking woman's attractiveness. Instead, the regular Jane is judged as plainer than if she had been seen on her own (without the prelude of beauties). In other words, the perceptual contrast effect happens similarly whether you are looking at members of your own sex or the opposite sex, and it is not connected to whether your mood goes up or down when you see the average person.

This disconnect between mood and judgment suggests that emotional reactions to other people and perceptual judgments of those same people are being calculated by two separate mechanisms in the brain. This directly contradicts the domain-general assumptions of the reinforcement-affect model. In other words, the one-principle theory, though parsimonious, is too simple.

There is another very influential domain-general theory of human relationships. Social psychologists call it economic-exchange theory, and it is a simple extension of what economists call utility theory, or the theory of rational man. According to this approach, rather than being like Pavlov's dogs, we are more calculating and rational in approaching our relationships. On this model, all human beings think about relationships in the same way that stockbrokers think about financial transactions—we buy in when it looks as if we will make a profit and sell if it looks as if we will take a loss. Whether we are thinking about friends, relatives, lovers, coworkers, or strangers, the general assumption is that we seek to optimize the ratio of costs to benefits. I'll talk about the economic view of man in more detail in Chapter 11, "Deep Rationality and Evolutionary Economics." For now I'll just point out that, as in the case of reinforcement-affect theory, economic theories have traditionally failed to address why the same outcome (a request for a kiss, for lunch, or for help writing a report) may sometimes be welcomed as a benefit and sometimes be regarded instead as a cost, depending on who is asking.

Multiple Minds

So the parsimony of domain-general theories comes at a price. They are often too lean on the devilish details that make for complete explanations. When I said earlier that the human brain uses different sets of rules for making decisions about the different people in our lives, I was instead advocating a *domain-specific* theory. In this view, there is no unitary "self" inside your head. Instead, there is a confederation of modular subselves, each one specialized to do one thing well.

Some of the best evidence for domain-specificity comes from research on animal learning, the same body of research that spawned reinforcement-affect theory. Running contrary to the vogue for simple models during the 1960s, several behavioral psychologists began to uncover evidence that the rules of conditioning seemed to change depending on what was being learned and which species was doing the learning. Consider the rules involved in learning to avoid poisonous foods. A fundamental behaviorist principle was that conditioning works best when there is instantaneous feedback. For example, if you receive a jolt of pain immediately after touching a hot frying pan, that teaches you not to touch a smoldering frying pan again; the pain does not get associated with the washcloth you used to wipe the pan out five minutes earlier. But research on conditioned nausea showed that the principle is sometimes violated. For example, one night I was hit with a dreadful bout of nausea at four in the morning. I did not associate the nausea with my bathroom or even with the Alka-Seltzer I had taken minutes before I became sick. Instead, I associated my upset stomach with a shrimp dish I had eaten at a party several hours before becoming ill (wrongly so, it turns out, because I was actually coming down with a flu). In this case, my mental mistake was based on a special exception to the rules of conditioning, an exception also found in other omnivorous animals. John Garcia and Robert Koelling found that rats learn to avoid foods that made them sick many hours after the food was eaten. Garcia and Koelling also found that food aversions are unlike many other types of learning in other ways: Food aversions require only one trial to learn and are very difficult to extinguish. In my case, for example, the powerful automatic aversion to possible toxins has totally overridden my conscious understanding that it was the flu and not the shrimp that was to blame—I still feel sick just thinking about those innocent little crustaceans. Garcia and Koelling's findings were revolutionary, suggesting that different forms of learning reflect different adaptive features. By conditioning strong aversions to any novel foods eaten several hours before, an omnivorous animal is better able to avoid toxic substances whose effects might not show up immediately. Like many novel ideas, though, Garcia and Koelling's findings were initially rejected by other scientists.

There is another functional twist to the research on food aversion: It is not just any cues that get conditioned to nausea. Instead, an animal's particular evolutionary history determines which associations that animal learns. For example, rats, which have poor vision and rely on taste and smell to find food at night, easily develop aversions to food that tastes unusual but not to food that looks unusual. Another team of researchers demonstrated quite a different pattern of learning for quail. These birds search for food relying not on taste but on their keen visual sense, and so quail condition nausea more easily to the color of a new food than to its taste.

So instead of having brains that operate on one or two domain-general rules, animals have subdivided minds. This revolutionary insight does not apply only to food aversions. David Sherry and Daniel Schacter reviewed numerous findings to show that there are several different kinds of memory, even in a bird's seemingly simple little brain. The rules governing how a bird remembers to tweet its species song are completely different from the rules that govern how it remembers where it stashed the winter supply of nuts, and both these memory systems are completely different from the memory systems that govern how it learns which foods are nutritious and which make it sick. Most importantly, the nature of the divisions inside any animal's head depends on the animal in question. Bats, for example, have a whole set of special mechanisms in their brains designed to allow them to paint a sonogram of the night sky. Their earthbound mammalian cousins, like us, do not.

What Does All This Imply for Me, Myself, and I?

It was fashionable for much of the twentieth century to argue that as the human brain grew larger, it lost most of its biological constraints.

But like tie-dyed bell-bottom trousers and smoking fat cigars in California restaurants, that argument has now gone out of fashion. For one thing, we humans, like rats, have different systems for remembering food aversions as opposed to food preferences. Our brains also have different systems for perceiving the sound of a sparrow singing and the color of its wings, and those systems are themselves subdivided (for example, the mechanisms for processing color involve different groups of neurons from those involved in processing shape and still different systems from those used to process movement). Likewise, our brains use one system for understanding the words we hear and another one for producing the words that come out of our own mouths.

Some of these different mental systems are geographically segregated, but some are not; many of them share some subprograms; and virtually all of them require appropriate developmental inputs to fully unfold (think of the simple two-word sentences coming out of a two-year-old's mouth and the complex things the same child can say two years later). Furthermore, there is enough neural plasticity that people who suffer brain damage can recover some of their lost abilities. But to say that the brain has some flexibility and that the brain's development involves continual interaction with the environment should not be taken to say that our heads contain blank slates full of interchangeable multipurpose neurons. Instead, our brains come with abundant preprogramming, which develops into a number of special suborgans during normal development.

How many subdivisions does the normal human brain have? We are still learning the answer to that question. But we now know that "one" was the wrong answer. Many evolutionary psychologists advocate what is called the "massive modularity" position. On that view, our heads are chock-full of independently operating little machines, each solving one particular problem. Researchers who have studied human emotions and animal instincts like to paint with a broader brush and think about organized systems of adaptive mechanisms. I think the two perspectives need not be mutually exclusive—that is, both may be true, depending on how you slice it.

I will return later to the question of how to subdivide the human brain after I talk about two relevant sets of research studies. One looks at some intriguing similarities in the sexual preferences of homosexual and heterosexual men. The other looks at the different reactions people have to sharing credit versus sharing sex with kin, friends, and strangers.

Homosexuality and the Modular Mind

When I was nineteen years old, the sexual revolution was raging. In New York City, where I grew up, there were love-ins in Central Park, and beautiful young women walked around in loose halter tops without bras, apparently proud of their newly liberated sexuality. One night I took the subway into Greenwich Village, dressed in my finest bell-bottoms and a peacoat, with the brilliant idea that I would hang out on the street until one of these sexually free beauties started flirting with me. I stood around awkwardly for a while, not getting much eye contact from the members of my target audience. Just when I was about to give up my plan, however, I heard the classic pick-up line: "Don't I know you?" However, it was not a braless young hippie woman with flowing blonde hair who was approaching me but, rather, a middle-aged black man in a conservative suit.

It turned out that I actually had briefly met this fellow earlier, so I did not know what to think when he seemed surprised by my "yes" answer. After a brief conversation, I realized that his question had only been a pick-up line and that he had forgotten an earlier conversation we had had when he came by the hotel where I worked as a doorman. I did, though, have a pleasant and illuminating conversation with the fellow, who had studied psychology in graduate school.

When I told him about my evening's plan to meet a woman, he wisely informed me that standing by myself on a street in Greenwich Village was more likely to result in meeting a homosexual man out for a one-night stand. I also had a brief chat with him about a troubling experience in which a man renting a room from my old girlfriend had made an aggressive pass at me, and I confessed that I was beginning to wonder if my attempt to act like a hip urban intellectual type wasn't somehow sending off incorrect signals about my sexual orientation. He reassured me that this was not the case, but he told me that homosexual guys were, like heterosexual guys, more proactive than women when it came to searching for new mating opportunities. Although neither of our sexual fantasies were fulfilled that evening, I did board the subway home with a slightly better understanding of the commonalities between heterosexual and homosexual men, as well as an unanswered question about why older homosexual men are interested in meeting guys so much younger than themselves.

A few years later, I actually did some research to help answer that question. As I discussed earlier, Rich Keefe and I had found that throughout their lives, except during adolescence, heterosexual men are interested in younger women, and that this desire seems to be all about fertility cues. And as I also discussed, women pursue older men because age in a man carries implications of status, wealth, and other advantages. An interesting question—and one that has direct bearing on our discussion of the modularity of the mind—is, What kind of men are *homosexual* men attracted to? As it turns out, homosexual men share with heterosexual men not only a tendency to be more proactive in seeking partners but also a host of other similarities. On the other hand, the guys homosexual men desire are surprisingly unlike those desired by heterosexual women. For example, Michael Bailey and his colleagues found that, like heterosexual men, homosexual men want a good-looking partner and do not care much about whether the guy is wealthy or high status. And when Keefe and I did a study of age preferences in homosexuals, we found that older gay men are, like older heterosexual men, attracted to much younger partners. This poses something of a problem for older homosexual men. Although an older heterosexual man stands a chance of finding a younger woman who reciprocates his interest, younger gay guys are not attracted to older men. Instead, like the older men, young homosexual men are interested in young men too.

The whole pattern is puzzling in several ways. Homosexual men's attraction toward younger attractive partners is not the result of reinforcement patterns (younger gay men do not reward an older guy's interest); it is not the result of adopting societal values about what is attractive in a man (or else gay guys and straight women would want the same thing); it is not the result of conscious rational decision-making (older homosexual men without permanent relationships often complain bitterly of loneliness, which could be alleviated by pairing up with another lonely older gay guy).

Homosexuals' general preference for members of their own sex still presents something of a puzzle from an evolutionary perspective, but homosexual men's preferences for young attractive partners are less puzzling. In fact, those preferences help shed additional light on the extent to which the brain is modular. The many similarities between gay men and heterosexual men suggest that human mating behavior, like human vision, is not simply a one-switch mechanism. Although the switch for sexual orientation has a different setting in homosexual and heterosexual men, for whatever reason, homosexual men's full pattern of preferences indicates that most of the other switches are still set at the same default settings as those of heterosexual men—generally speaking, both want young attractive partners, both want a lot of partners, and neither cares about the wealth or status of their partners. In most ways, other than the direction of their preferences, homosexual men are playing out a mating strategy that would otherwise result in high reproductive success (if their targets were women).

Is Friendship Akin to Kinship?

On our European vacation from hell, I tolerated a lot more annoyance from my son Dave than I did from my lifelong friend Rich. Indeed, the way people feel about their friends as opposed to their relatives also offers some interesting insights into our minds' modularity. In some ways, friends and relatives share a similar place in our hearts, both typically triggering a lot of positive associations. My former student Josh Ackerman and I conducted an experiment in which we asked subjects to play a team quiz game, either with a parent, a sibling, a friend, or a stranger. All the teams were told they had done exceptionally well, regardless of their actual performance. When we asked them who was responsible for their team's success, people playing with strangers took the lion's share of the credit. People who played the game with relatives gave more credit to those family members. Friends were generally given more credit than strangers, although there was an unexpected split along sex lines. Women shared credit with their friends as though they were kin, but men treated friends a bit more like strangers.

Working with Mark Schaller, Josh and I did another study that moved into a rather taboo area at the borderline between friendship and kinship. We asked college students to concentrate on what it would be like to have sex with a stranger, a friend, or a close relative. It may surprise you to know that we are not the only researchers who have asked people to imagine having sex with relatives. In fact, consistent with their inclination to have one foot in the gutter, several evolutionary psychologists—including anthropologist Dan Fessler and psychologist Deb Lieberman—have done similar work, and we have all found similar results. People, both men and women, generally find the very thought of sibling sex quite disgusting. On the other hand, when thinking about sex with strangers—people who are not usually the targets of positive associations—our subjects experienced almost no disgust and more positive than negative feelings.

Psychologist Lisa DeBruine approached the question in a different way. She used computer morphing software to make a stranger look like a relative (by blending an image of a subject's face with an image of an opposite-sexed stranger's face). She found that people judged the faces with artificial kinship cues as "trustworthy, but not lustworthy." That is, making another face resemble yours increases the odds that you will see this person as someone on whom you can depend but not as someone you would desire for a short-term sexual liaison.

At one level, people ought to be especially attracted to their close relatives (who otherwise meet many of the criteria for desirable mates, such as being very similar and very familiar). Why does the very thought disgust people? From a genetic perspective, it is as close to cloning ourselves as we could get. At first blush, it seems as if anything approximating cloning would serve a gene's interests (by maximizing the genetic overlap between parent and offspring). But something gets lost with too much togetherness. One key advantage of sexual reproduction is that it shuffles our genes with another set, which helps keep ahead of all those rapidly evolving viral and bacterial parasites. Besides insufficient shuffling, mating with first-degree relatives results in what animal breeders call "inbreeding depression." What that means, biologically speaking, is an increased chance of combining rare recessive genes that underlie harmful genetic disorders.

What happened when people thought about sex with friends threw us another interesting conceptual curveball. Men thinking about sex with their friends responded as they had to thinking about sex with strangers: Their positive feelings strongly outweighed their negative feelings. Women, on the other hand, reported slightly more negative than positive feelings, with disgust the most prominent reaction. To raise this to a higher intellectual plane, if the *Friends* television characters Joey and Monica were to imagine having sex with one another, our results suggest that the thought might be appealing to Joey but slightly disgusting to Monica.

These results suggest two things. First, our brains do not operate according to one simple rule when it comes to thinking about people we regard as "attractive" or "likable." Both men and women keep kin in a separate mental category from attractive strangers. Second, when thinking about friends, men and women do not use the same decision rules. In some ways, women treat friends like kin, whereas men treat them like strangers. But even that is an oversimplification. Although men often compete with male friends and feel sexual attraction toward female friends, they do not use the same rules for sharing rewards with friends as with strangers, and they are more likely to trust friends than strangers. And women are not nearly as disgusted by thoughts of sex with friends as they are by thoughts of sex with relatives. In other words, our brains use separate accounting systems for relatives, friends, and strangers.

Evolutionary theorists make a key theoretical distinction between kin and nonrelated friends. Because relatives share a higher percentage of identical genes with one another, anything we do to benefit a relative's reproductive success indirectly benefits our own fitness. The technical term that evolutionary biologists use is "inclusive fitness"— which refers broadly to an individual's success at passing his or her genes on. My contributions to my son's reproductive success count toward my own, as would anything I might do to help my sisters. Inclusive fitness helps explain why I was a lot more tolerant of my son's European-vacation teenage whining than my ex-wife was (she was not his mother).

Helping between friends, on the other hand, is typically explained in terms of reciprocal altruism—helping that continues as long as you keep scratching my back at about the same rate as I scratch yours. Reciprocal altruism is a powerful rule: It allows humans in groups to accomplish many things they could not accomplish alone, and when times are tough, it may mean the difference between survival and starvation. But it works on a slightly different math than inclusive fitness. Every time I give something to my son I give something to my own genes; that bond is always there. The same does not hold for my friend Rich. If he and I stop getting rewards from our mutual interactions, as we did during our ill-fated European vacation, the bond can be threatened. Although Rich and I had a long enough history of mutual reward that we were able to get over a few bumpy weeks, some relationships end forever when former friends calculate an unfavorable answer to the question, "What's in it for *moi*?" As for the French bakerwoman who was rude to me for a couple of minutes, she and I had neither type of bond going, so when our short relationship started so negatively, it cost me nothing to move her over into the "enemy" category.

So How Many Subselves Live Inside Your Head?

In the first psychology textbook ever written, William James suggested that "a man . . . has as many social selves as there are distinct groups of persons about whose opinion he cares." James's suggestion fits with the research I just discussed—perhaps we have different functional subselves for dealing with different categories of people: relatives, friends, potential mates, romantic partners, ingroup acquaintances, and enemies. That model beats the alternative view that we have only one single tightly integrated self, but it has a key limitation. Sometimes the same other person can be a friend, a competitor, or a sexual opportunity, depending on other factors in the current situation.

I think we get a better answer to the question, "How many subselves?" if we first ask another: "What are the key sets of problems human beings typically have had to solve?" That would suggest a different subself for dealing with qualitatively different situations: dangers of violence, disease, or losing romantic partners; or chances to find new mates, gain status, establish friendships, or provide for our close relatives.

The idea of functional subselves has guided my team's research on what we call "fundamental motives." When you are under the influence of a different fundamental motive, such as mating or self-protection, you are a different person—you notice different things and you remember different things, and that leads you to respond differently to the same situation. I have already talked about how activating different motivational subselves linked to mating, self-protection, or disease avoidance can lead to different patterns of prejudice and to different responses to a member of the opposite sex. In the next few chapters, I will talk more about how the different biases associated with each of these subselves can influence our inclinations to spend money on flashy goods, to conform to group opinion, or to want to go to church.

Here is a list of the separate characters that I think you and I have running around inside our heads, and what set of problems or opportunities each of those subselves is in charge of managing:

- *The team player:* One of our subselves manages problems and opportunities related to affiliation. To survive and reproduce, our ancestors needed to get along with other people. Friends share food, teach us valuable skills, and fill us in on essential information; they team up with us to move things that are too big; and they provide safety in numbers when the bad guys are around. But there are costs to friendship. Sometimes friends take more than they give, and insiders are in the best position to betray us. The team-player subself is tuned in to information about which of our acquaintances might make good friends, whether we are being accepted or rejected by those people, and whether we are getting along with our old friends.

- *The go-getter:* Another subself manages problems and opportunities related to status. Being respected by others brings numerous survival and reproductive benefits; being disdained carries some serious costs. But respect and status do not come for free: Leaders have to give the group more than followers do, and people do not like it when their friends step over them. The go-getter subself is tuned in to where we stand in the dominance hierarchy and to who is above and below us.

- *The night watchman:* This subself manages problems and opportunities linked to self-protection. The night watchman subself is tuned in to information such as: Is that band of nasty looking guys who just walked over the hill going to steal something from me or burn down my hut? Are there enough of my tribe members around that I can protect myself?

- *The compulsive:* This subself is in charge of avoiding disease. Why is that other person coughing? Does something smell rotten in here? Should I wash off after shaking hands with that stranger?

- *The swinging single:* This is the subself concerned with acquiring mates. As I have discussed, the "his" and "hers" versions are somewhat different, tuned to the sex-specific cues that make for good mates.

- *The good spouse:* This subself is in charge of retaining mates. It is tuned in to information about whether my partner seems to be happy or unhappy, and it is also scanning the social horizon for potential interlopers who might be in the market to make my partner happier.

- *The parent:* The parental subself manages threats and opportunities linked to kin care. It is responsive to information about whether one's children, grandchildren, nephews, and nieces are doing well.

At any given moment of consciousness, only one of these subselves is running the show. When you are worried about the band of knife-wielding thugs who just walked around the corner, you are not thinking about romancing your date. Some of your subselves have common goals—befriending a neighbor could simultaneously serve affiliative, self-protective, and parenting goals, for example. But some of them have incompatible goals—your swinging-single subself and your good spouse subself being the most obvious example.

These different subselves, in some sense, serve as the packages for the modules I was discussing earlier. The data on homosexual men's choices, for example, suggest a lot of separate mental switches linked to the mating motive, not just one. Or, to take another example, if you trigger a mechanism designed to recognize an angry face, it activates another set of linked mechanisms involving self-protection. That said, most of our mental modules are not rigidly encapsulated—that is, they share some software and hardware—as when women treat kin and friends using many of the same decision mechanisms. In this sense, your mind is very much like a computer: It has different named programs on it, but all of them rely on the same hardware for inputs and processing and use much of the same core code of the operating system to get their tasks done.

In the next chapter I will describe how different subselves come on line at different stages of our lives, and in Chapter 8 I will talk about how our mental processes change radically depending on which of our different subselves is currently in charge.

Subselves on Vacation

Thinking about subselves helps me understand why things were so stressful on my European vacation. It is easier to make decisions when there is only one subself in the driver's seat. At any given moment, in fact, there can be only one driver—my body can only walk in one direction and my conscious mind can only think about a limited amount of information. During the portion of the vacation when I was biking alone with my son, my parental subself got to drive for most of the day, and it was a lot less exhausting. If I had been traveling alone with my friend Rich, then my affiliative subself would have been in the driver's seat, and it would have gone more easily. In fact, Rich and I took a trip to Mexico a decade later, without our families, and it was quite pleasant. Likewise for travel with Melanie; she and I traveled to Italy alone on another junket and it went splendidly. But when we were all together, and my parent subself was pitted against my spouse subself and my affiliative subself was pushing in another direction, it got exhausting. And to make matters worse, my slightly paranoid night watchman subself is often on alert when I am traveling, ever attentive to the possibility that the next stranger could be a thief, a pickpocket, a mugger, a terrorist, or—worse yet—a French bakerwoman.

Chapter 7

RECONSTRUCTING MASLOW'S PYRAMID

It's 1967, and I have a summer job working 4:00 p.m. to midnight as a doorman at the Paramount Hotel in New York's theater district. On a typical night, I go through all the steps in Abraham Maslow's famous pyramid of motives. By dinner break, I've been carrying luggage and chasing down cabs for several hours, and I'm starving and thirsty. So I order a couple of slices of pizza and a chocolate malt from the Greek joint around the corner, to the sound of the Young Rascals' "Groovin'" on the radio back in the kitchen. Returning from my meal, I worry about my physical safety as I pass a nasty looking junkie on Eighth Avenue. Safe and sound back at the hotel, I strike up a friendly conversation with a college student from Georgia who is visiting New York for the first time. After nine or ten in the evening, business slows down, and I get to finish off my shift reading psychology books, as I dream about someday becoming a university professor and fulfilling myself as an intellectual.

It was during my doorman period that I first came across Maslow's pyramid (Figure 7.1). Maslow's powerful visual image of a pyramid of needs has been one of the most cognitively contagious ideas in the behavioral sciences, so you've seen this picture if you've ever taken a

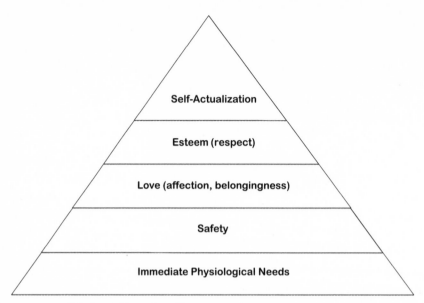

FIGURE 7.1 Maslow's hierarchy of human motives

course in psychology or organizational behavior. Maslow was spurred to develop his model of human motivation because he disagreed with the prevailing paradigm in psychology at the time. As I discussed in Chapter 6, behaviorists—who dominated psychology in Maslow's day—tended to believe that all human motives could ultimately be traced back to a few "primary drives" such as hunger and thirst. On that view, my young son seeks contact with his mother because he has learned to associate that contact with a primary reward—food. Psychologists at the time likewise viewed adults' desires for love, respect, and intellectual fulfillment as "secondary drives"—desires rooted in past experiences involving their parents rewarding them with milk and cookies when they acted affectionately, hit a home run, or got a good grade in school.

But Maslow did not buy the argument about secondary drives. He argued, "We could never understand fully the need for love no matter how much we might know about the hunger drive." He proposed instead that human beings have several completely independent sets of basic motives. At the base of his hierarchy were physiological needs like hunger and thirst. These are biological priorities because you would die if you ignored them (that pizza was my top priority after a few hours of carrying people's luggage). If physical needs like hunger are satisfied, though, people naturally move on to worrying about safety—you will risk your life if you are about to starve, but once you are reasonably well fed, you cover your ass. (Maslow would have predicted that I would have been more likely to worry about the mean-looking junkie on the way back from my dinner than on the way there.) Next in the hierarchy are social motives—the desire to have some friends who will show you affection, followed by the desire to get some respect. (I'm fed, I'm safe, and now I have time to plug into the social network.)

At the pinnacle of the pyramid of motives was *self-actualization*—the desire to fulfill your unique potential, as a musician, a poet, a philosopher, or whatever you happen to excel at. Right now I am locked away in my summer house, ignoring the sounds of the goldfinches and chickadees that tempt me to go for a hike in the beautiful mountains, suppressing my desire to drive into town and eat lunch at my favorite Mexican restaurant, and avoiding any other contact with my fellow human beings. What I am doing instead is writing a book about sex, murder, and the meaning of life, pondering puzzling life experiences and big ideas while trying to integrate my writing and teaching abilities at their highest level—just the kind of thing I dreamed about doing when I was working as a summer doorman at the Paramount.

Sex and the Meaning of Life

Maslow's ideas were avant-garde at the time, but several of them have been solidly supported by later research by neuroscientists and evolutionary psychologists. Maslow guessed that people everywhere shared a set of universal fundamental motives, and as I have already discussed in this book, he was right. Maslow guessed that human brains did not operate according to one simple set of rules but, instead, used different subsystems to accomplish different goals. There is now plenty of evidence that he was right there, too. Today we would call this the "modularity" hypothesis, one of the foundational ideas linking modern cognitive science and evolutionary biology. And although the evidence is not yet all in, Maslow's assumption that some motives take priority over others was probably also correct.

But we have learned a lot since Maslow's day, and although his pyramid is worth preserving, it needs some remodeling to bring it up to twenty-first-century building codes. With a bit of reconstruction, though, this skyward-pointing edifice can encompass the different building blocks from our earlier chapters. Indeed, a rebuilt pyramid can help us ascend from the gutter to the stars by clarifying the structural connections that link sex and murder with conspicuous consumption, art, religion, and the meaning of life.

Probably the biggest problem with Maslow's pyramid is this: He did not understand the central importance of reproduction to human life. Maslow did mention sex occasionally in his writings, but he discussed it mostly as a simple physiological need—an irritation we need to scratch before moving on to bigger and better things (like playing classical guitar or writing poetry). And Maslow had very little to say about other aspects of human reproduction, such as taking care of the children likely to follow from scratching our sexual itches.

Maslow's pyramid also had another basic problem in its blueprint. He believed that the motives at the top of the hierarchy are somehow disconnected from biology. In fact, he later called the lower needs "deficiency needs"—linking them to basic biological processes that keep us alive by avoiding starvation, bodily harm, and being cast out of our social groups. He placed intellectual curiosity and self-actualization on a higher plane, above biology, distinguishing them as "being needs" or "growth needs." This distinction flowed logically from his idea that previous psychologists had failed to distinguish humans from other animals. And at first glance, this logic seems reasonable: We share needs like hunger, thirst, and self-protection with other animals, but animals do not share our need to write poetry, play music, study philosophy, and build structures like the Egyptian pyramids.

There is actually a logical error in that line of reasoning, but the mistake is so cognitively compelling that many intelligent people find it hard to resist to this day (and it lurks behind some of the opposition to evolutionary psychology). Does the fact that other animals do not appear to write poems, compose music, or study architecture prove that our cultural inclinations are somehow on a different plane from normal biology? If you take our exalted selves out of the picture, you can see the problem more clearly: Most animals do not have the ability to create ultrasound images of the night world in their brains, as bats do, and most animals do not have the ability to navigate using polarized light, as bees do, but we do not place the unique characteristics of bats and bees in a special "nonbiological" category. Nor should we be so quick to do so for behaviors that seem to be uniquely human. As I will discuss in Chapter 9, there is now evidence that the highest realms of human creative genius are intimately connected to fundamental biological processes—processes that link us directly with the rest of the biological world. And as we will see, the same holds for a motivation that humanists literally and figuratively hold sacred: the inclination toward religiosity.

Over the last few years, with a team of brilliant colleagues and students, I have been conducting research on fundamental human motives, including many of the findings I have been discussing in this book. Based on that program of research and on the theoretical advances in evolutionary biology and cognitive science that underlie it, I teamed up with Vlad Griskevicius, Steve Neuberg, and Mark Schaller to develop a blueprint for a reconstructed hierarchy of motives, which you can see in Figure 7.2.

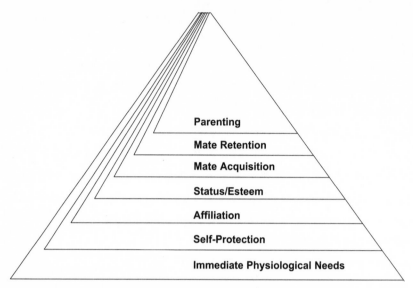

FIGURE 7.2 An updated hierarchy of fundamental human motives

There are three important differences between our new hierarchy and the old one. First, self-actualization is displaced from its hallowed position at the top of the pyramid. I am not saying that people do not experience all those "higher" strivings; when I play classical guitar, write creative essays, or try my hand at using watercolors on a canvas, it feels as if I am doing it for the sheer pleasure of creating. But if we think in terms of evolutionary function, a big part of what Maslow meant by self-actualization (including poets' and artists' and musicians' striving for perfection) can be neatly folded into the esteem category. Throughout history, people who perfected their creative performances or showed off their intellectual capacities often gained status, and that often improved their odds of reproducing. I realize that to many humanists that is the equivalent of saying there is no Santa Claus, but the fact of the matter is, although gifts do arrive under the Christmas tree, they were delivered there by a very nonmystical process, one involving a less jolly mortal driving a station wagon to Toys"R"Us with a credit card at the ready. Likewise, the motive to self-actualize is based in more mundane strivings.

Second, there are now three new motives at the top of the hierarchy, all linked to reproduction. This change follows from a powerful biological theory of development called life history theory, which we have already briefly encountered but which I will describe in more detail below. Notice that none of the new goals is actually called the "desire for sex." That desire is there, moved out of Maslow's lowest category of physiological needs and up into the mate-acquisition category. In some ways, sexual desire plays a minor role even there, because scratching the copulatory itch only happens after people have expended a great deal of time and effort finding and choosing the right person with whom to scratch it. Nevertheless, despite the fact that most people (who aren't porno stars) spend very little time on sexual intercourse per se, it is essential that we get around to it at some point. We are, after all, a sexually reproducing species, and our genes are not going anywhere without it.

After finding a satisfactory mate and convincing that person that you are also good enough to consider as a scratching collaborator, a whole new set of goals arises—maintaining that relationship. Sticking together with a partner involves a completely different set of problems from those involved in finding one, and some people (including me at several phases of my development) can be pretty good at the finding part but not so good at the staying together part.

At the top of the new hierarchy is parenting, which involves still another set of goals beyond finding a mate or staying in a romantic relationship. Romantic couples often experience a sharp drop in affection for one another after the arrival of a child, but they are nevertheless more likely to stay together after the new, and typically quite demanding, family member comes between them. *Homo sapiens* is in fact one of a small minority of mammalian species in which the father often sticks around to contribute to the offspring—and fatherly love is an inclination that coevolved with the development of big-brained but helpless offspring who require tremendous doses of care and nutrition. The human parental bond is critically important; children in traditional societies are much less likely to survive without two parents working together to meet their seemingly incessant needs. Again, it might seem sacrilegious to replace self-actualization with something so mundane as reproduction, but I will argue later that not only is this change more scientifically correct, but it also puts a more positive spin on human nature.

There is a third difference between the new hierarchy and the old. Rather than stacking the goals on top of one another, the goals in the new hierarchy are overlapping. This graphic twist is meant to capture another important point: The higher needs develop later, but they do not replace the lower ones. Instead, our earlier-developing needs remain centrally important, receding into the background only until they are triggered by opportunities and threats in the environment. As part of his reaction against radical behaviorism (which completely denied the causal role of factors inside the person), Maslow liked to downplay the role of the environment. He also liked to argue that those self-actualizing poets, artists, and musicians, when they were at their best, had mostly risen above all their less exalted concerns over food, safety, friendship, and esteem. But again, at the risk of saying Grinch-like things about Santa Claus, it ain't quite like that down in Who-ville.

The Evolutionary Functions of Fundamental Motives

So why do we need love, esteem, pizza, and personal fulfillment? Let us first try to answer that question in terms of evolutionary function. At one level, everything any animal does is the product of mechanisms that enhanced what biologists call inclusive fitness (as I described earlier, that term refers to success at assisting one's genes on their trip into the future). But neither you, me, nor your pet parakeet experiences any general motive to "increase my inclusive fitness." Instead, we have toolboxes full of lower-level motives, each designed to solve a different set of life problems.

What Maslow called physiological needs, such as hunger and thirst, are clearly designed to help us survive. As Maslow pointed out, there are lots of specific needs. Even hunger can be subdivided into particular kinds of cravings for particular nutrients at particular times when we most need them. During pregnancy, for example, women get particular cravings for certain foods, but they get violently ill at the very thought of others. Pregnant women steer clear of vegetables containing high levels of toxins and foods likely to carry bacteria (such as fish), especially during the first trimester, when the growing fetus is especially sensitive.

Safety needs also serve an obvious and essential survival function— protecting us from potential harm. Research by Arne Öhman and his colleagues has shown that the fear system, designed to rapidly respond to threats, comes preequipped with some of its own adaptively tuned wiring, just as the hunger system does. For example, it is difficult to condition a fear response to flowers or abstract paintings, but it is very easy to learn to fear a dog or a snake.

The benefits of social belongingness are not as direct and immediate as those for hunger and fear, but they are there. Our ancestors would have had a very hard time surviving without a network of friends. Evolutionary anthropologists Kim Hill and Magdalena Hurtado studied the Aché in Paraguay, who live in conditions much closer to those of our hunter-gatherer ancestors. They found that tribe members shared food in a way that provided an insurance policy against starvation: I may bring home a wild pig this week but then have several weeks of bad luck; my neighbor may get lucky on a different week. If we share, we pool our risk.

To facilitate our alliances with other people, we seem to have evolved yet another set of hormonal and neurological systems. But here Maslow did not go far enough in his modularity. He lumped romantic love, bonds between family members, and affiliation with groups into the same category, when in fact there are important functional and neurological differences among the biological systems that manage different kinds of relationships. Our dealings with family members, romantic partners, and friends are governed by distinct emotions and cognitive biases. For example, sexual arousal and sexual jealousy are distinctly designed to deal with the opportunities and threats that arise in romantic relationships, but not in relationships between children and parents. As another example, if a friend left me to clean up after I bought his food and prepared his dinner every day and then asked me to pay for his college education and buy him a car, I would quickly start looking for a new friend. But in my relationship with my two sons, I take it for granted that we will have just such a grossly uneven distribution of benefits and costs. And there is another reason to separate familial bonds, friendship, and romantic love in our hierarchy: They come to the fore at very different developmental periods.

What are the functional benefits of esteem needs? Getting others to hold you in high esteem translates into status in your group, and that translates into numerous benefits, such as privileged access to the watering hole and the raspberry bush. If you are a female, it means more goodies to pass on to your offspring. If you are a male, there is an additional benefit, for, as we have seen in earlier chapters, males with high status get access to more females; males with very low status may not get a mate at all. Hence all that showing off by males: painting thousands of artworks, like Pablo Picasso or Diego Rivera; writing volumes of poetry, like Pablo Neruda; composing catchy melodies, like John Lennon or Duke Ellington; or doing whatever else one happens to be good at.

Maslow's view of self-actualization was that it had no biological function. If we view self-actualization as an extension of esteem needs, though, it has a clear function. If you perfect your music, art, or writing abilities, you can win respect and admiration from others, and all the benefits that come with them.

Life History Theory and the Developmental Hierarchy

Besides his idea that some motives take mental priority over others, Maslow's theory had a developmental component. He assumed that our priorities shift over the life span, that we move up in the hierarchy as we mature. For example, when my younger son was an infant, he was only concerned with physiological needs. He wanted to be nursed, he wanted his diapers changed, and he was quite willing to scream until he got some service. He did not start worrying too much about affection until a few years later. Around the time he entered kindergarten, he began to surprise us by expressing a sudden interest in what other people wanted. And it was not until he made some friends that he began to worry about whether they respected him or not.

Maslow's theory sometimes blurred together cognitive priority (what takes precedence in our minds) and developmental priority (what comes first in our life spans). He presumed that the two kinds of priorities more or less move in synchrony with one another. But they may not. Physiological needs like hunger and thirst do in fact tend to show up early in development, as in the case of my hungry son. But other physiological needs, such as the hormonally driven desire for sexual satisfaction, do not even come on line until adolescence. (When my older son was in grammar school, he reacted to a song line about "two girls for every guy" with a "yuck"; when he was a senior in high school, he could not believe he had ever said that.) Even as adults, though, we can suppress sexual desire, as well as other physiological needs such as hunger, if we think other people will disapprove. So the developmental hierarchy need not correspond precisely to the cognitive hierarchy. Moreover, our cognitive hierarchies might change at any point in our lives, depending on where we are and who is around. For example, when my older son is writing a screenplay and his artistic subself is in charge, he can go for hours ignoring physiological needs that would monopolize an infant's attention, but only up to a point—he always stops for a meal in time to avoid starvation. So a person's currently conscious priorities and the order in which fundamental motives develop are two issues worth keeping separate.

Biological theorists have developed a powerful set of ideas called life history theory, which has profound implications for thinking about the development of human motives. And those ideas explain not only why there are developmental priorities at the bottom of Maslow's hierarchy, but also why we have rebuilt the top sections.

Life history researchers try to answer questions like, Why do some creatures spend longer or shorter periods developing their bodies before they start reproducing? For example, one species of tenrec (a small mammal found in Madagascar) starts reproducing five weeks after it is born. Elephants, on the other hand, take decades to reach sexual maturity. Once an animal matures, does it devote all its resources to one short reproductive burst, like a salmon, or does it spread those reproductive efforts over several episodes spanning months or years, like a tortoise? Does the animal allocate resources to caring for its offspring after they are born, and if so, how much care does it invest before leaving the offspring to fend for themselves? Some fish spray their eggs on a river bottom, for example, and that is the end of family togetherness. Not only do human beings help their children for decades, but they often help raise their grandchildren.

Life history theory is at heart a theory about biological economics. A central assumption of the theory is that every animal has a limited budget of resources. So as the animal develops, there are always trade-offs in when and how to allocate those scarce resources. Life histories can be divided into two major phases, each involving different trade-offs. The first phase focuses on somatic effort, which refers to the energy every animal must expend to build its body (in economic terms, it is like making investments in a biological bank account). The later emphasizes reproductive effort (spending that bank account in ways that will replicate the individual's genes). Reproductive effort for some animals, like us, can be further divided into mating and parental care.

As in Maslow's pyramid, life history tasks fall into a natural hierarchy. Somatic efforts are at the base, because an animal cannot mate until it builds an adult body, and it cannot invest in parenting until it finds a mate. In any species that reproduces more than once, as we do, these goal systems do not replace one another. Adult animals need to divide their current resources among somatic effort (eating, drinking, and protecting themselves), attracting and keeping mates, and caring for offspring. Given that resources are inherently finite, time and energy invested in one activity must be taken from others. Time spent looking for new mating opportunities is time not spent parenting.

Why don't all animals start reproducing as soon as they can, like tenrecs, and have as many offspring as possible? The answer is that the goal is to produce *viable* offspring, and that goal might not be accomplished if the animal produces too many too soon. What is an optimal investment of reproductive effort? That depends on the features of the particular species and the particular ecological constraints faced by that species. For large mammals like elephants, females are not physically capable of producing and nurturing offspring until they are several years old. And for elephants, as for any species providing parental care, having too many offspring too soon decreases the chances that any of them will survive.

Humans are nearer to elephants than to tenrecs in our developmental life histories. We naked apes do not reach sexual maturity for over a decade. We devote the long nonreproductive period to developing our bodies, and those bodies do not even bring secondary sexual organs on line until we need them. During that time, we are also feeding energy to our information-hungry brains as we learn the critical social skills that enable us to establish a network of friends and gain some respect from the other people in that network. Even after we reach sexual maturity, we do not just run off to the mating grounds, like other mammals do; we spend several years seeking a mate. Our ancestors would have then turned most of their energies to parenting, caring for those slow-maturing, large-brained offspring, who were much more likely to thrive if they had two parents looking after them.

It is these ideas from life history theory that inspired us to add three separate motivational systems to Maslow's hierarchy: mate acquisition, mate retention, and parenting. In the new pyramid, survival and social goals provide the foundation for acquiring mates. Getting a mate provides a foundation for forming a long-term bond, and this in turn undergirds the goal of producing offspring and then successfully raising them.

Evolutionism, Humanism, and Positive Psychology

Maslow distinguished his humanistic approach to psychology from psychoanalysis and behaviorism, the two other major perspectives of his time. He argued that the psychoanalytic approach was flawed in its obsession with the negative and pathological aspects of human behavior, viewing people as driven by suppressed feelings of hostility and sexual desires—often directed at their mothers. The behaviorist approach was limited, in Maslow's view, by its assumption that we can discover the general principles of human behavior by studying rats. Maslow believed that psychologists had ignored the many positive traits of humans (such as artistic creativity and scientific curiosity) because those were not to be found in rats or in people too depressed to get out of bed in the morning.

At first glance, it may seem that an evolutionary approach, à la Freud, adopts a view of humans as driven by base sexual and aggressive instincts and, à la Skinner, emphasizes the commonalities between humans and rats. But first glances can be deceiving. The modern evolutionary approach is actually compatible with the two points Maslow liked to emphasize: (1) Human psychology is different from that of other animals, and (2) psychologists need to understand the positive as well as the negative aspects of human behavior.

Although an evolutionary perspective recognizes sexual and competitive motivations as undeniable aspects of human nature, it also emphasizes the importance of cooperation, love, and parental concern for survival in human groups. Again, it is important to reiterate: Sex is only a small part of human reproduction. We human beings devote immense effort to long courtship periods, and even for the sexiest among us, courtship usually involves more hours spent in platonic activities than in copulation. After initial courtship, humans devote a great deal of energy to maintaining their bonds and raising their children. And for the last few decades of their lives, human beings may devote great quantities of energy to helping their grandchildren.

As I will describe in the next few chapters, reproductive goals are the ultimate driving force behind much that is positive in human nature, such as creating music and poetry, devoting oneself to charitable endeavors, or working to improve the world for the next generation. Developmental psychologists have found that as people age, they tend to become increasingly concerned with the welfare of other people. Thus, Maslow's idea of self-actualization (which often involves pursuing what brings personal pleasure) is just a self-centered step on the way to a much higher goal—taking care of other people.

So the renovated pyramid of motivation helps us see the tight linkages between topics as disparate as sex and self-actualization, clarifying the structural connections between the muddy bricks down by the gutter and the shinier ones up near the stars. And it points us to a higher meaning of life, lifting us above our immediate cravings and into the firmament of social interconnectedness.

There is one point on which Maslow, Freud, and Skinner agreed: that people are usually unaware of the underlying causes of their behavior. On this point, I will join the chorus. Remember the birds that migrate when the days start getting shorter; they have no idea of the connection between their urge for going and how it links up to finding food, nesting sites, and their ultimate reproductive success. Likewise, although people are good at making up explanations for their behaviors, we certainly do not consciously experience the links between those behaviors and their ultimate functional goals. At the functional level, though, everything we do is intimately linked together. Eating, drinking, and staying out of dangerous neighborhoods at night serve the higher goal of surviving long enough to mate. Playing nice with others and striving for their respect serve the higher goal of finding mates, and finding mates and trying to stay together with those mates serve the higher goal of having children. Taking care of the children serves the higher goal of increasing our inclusive fitness. Those connections are not conscious and they do not need to be, any more than the connections among day length, migration, and inclusive fitness are conscious in a scarlet tanager.

Proximate Motives: The Ebb and Flow of Our Latent Subselves

In Maslow's view, the "primitive" physiological needs, such as hunger, recede into the background in healthy growth-oriented adults. But is that really so? Think about hunger. Even relatively contented, well-connected, high-status people often devote a great deal of attention, conscious thought, and conversation to selecting, preparing, and presenting food. My wife has a Ph.D. and does research for a living, but her favorite reading material includes the recipes in *Cook's Illustrated* magazine. In fact, several of my most self-actualized friends say that they look forward to retirement not simply as a time to read and think more, but also as a time to spend more time cooking enjoyable meals.

The same holds for the later-developing needs, such as the need to belong. Adults, even attractive and well-connected college students, remain exquisitely sensitive to social acceptance and social rejection, and when they experience social isolation, the pain is registered by the same physiological mechanisms used to register physical pain. Research since Maslow's time supports a view that later-developed motivations *build upon rather than replace* earlier ones.

So although earlier-developing needs do have to time-share with the ones that come later, they do not disappear in healthy, well-functioning adults. Instead, as depicted in the overlapping triangles in the new pyramid, they remain in the background to respond as relevant threats and opportunities arise. This directly links up to the most proximate level of analysis—the way different goals are calibrated to what is currently going on outside in the world.

Maslow's tendency to downplay the environment was linked to a bias in the humanistic movement that his writings helped spawn. Humanistic psychologists sometimes adopted an almost solipsistic emphasis on personal phenomenology: Don't like your perceptions of the world, just change your thoughts. On one level, it sounds nice to contemplate your own belly button, think your own thoughts, and do your own thing. But it is ultimately not the way people are wired; we're just not that self-centered. And it is not actually a higher plane of existence to be disconnected from the people around us. Indeed, adults who do not pay any attention to other people's needs may be manifesting more pathology than self-realization.

As I discussed in Chapter 6, we are all multiple personalities, and each of our subselves is driven by a different motivational subsystem.

Which one of our subselves comes to the surface depends on what is going on in our current situation. When we are in the dark, we feel especially vulnerable to harm from strangers. As a consequence, our night watchman subself takes charge, and we are, as discussed in Chapter 4, more likely to perceive a man from an ethnic outgroup as aggressive and untrustworthy. In the next chapter, we will see how our different subselves warp our minds in very different ways depending on who is around and what is on our motivational agenda.

Chapter 8

HOW THE MIND WARPS

Although it was over thirty years ago, I remember my mother's exact words as she spoke to me over the phone: "Douglas, if you're standing, sit down. I have some very bad news for you." The news was about my little brother James, and it was very bad indeed. James was dead, crushed by the wheels of a train on the Long Island Rail Road. Although I have no memory of what I was doing before she called or what I had been doing any time earlier that week or that month, I do remember exactly where I was standing when I got the call: staring at a yellow wall in my little kitchen in Bozeman. And I even remember how the room was lit by the afternoon sunlight filtering in from the outside porch.

For this painful moment in my life, I have what social psychologists Roger Brown and James Kulik called a "flashbulb memory." Most Americans have a flashbulb memory of the moment they heard about the planes crashing into New York's World Trade Center—a precise picture in their heads of where they were and what was happening when they got the news. In trying to dredge up my own flashbulb memories, I uncovered a couple of unpleasant mental photographs in which something socially embarrassing happened to me. Although you might find it amusing to hear about the time I got plastered drunk at a conference and made a spectacle of myself, I must plead the Holden Caulfield defense: I don't want to talk about it, if you want the truth. As I kept searching my memory banks, though, I discovered that not all my flashbulb memories are negative. Indeed, some are rather delightful. I remember the precise setting and lighting conditions in the beautiful side canyon of Sedona's Oak Creek where I experienced a three-way kiss with two very attractive women. I might enjoy talking about what happened next, but in this case, you probably do not want to hear about it, and they probably do not want me to give any more details.

As I continued dredging my mind for flashbulb memories, though, I noticed some surprising blank spaces on the tapes. For example, I vividly recollect a pretty young blonde woman who smiled pleasantly at me as we boarded the same plane several decades ago. I remember exactly what she looked like, the color of her hair, and how she was dressed. I also remember thinking I might introduce myself to her, until I read the message on her shirt: "If you ain't riding a Harley, you ain't shit."

I had no doubt what category my dinky little Japanese car put me in, so I decided not to chat her up. Good thing, too, because I also vividly recollect the guy who picked her up at the airport when we landed: He was well over six feet tall, a solid 250 pounds, and adorned in a black leather jacket and black boots. He had numerous thick silver chains hanging from his black jeans, alongside what appeared to be a gun holster and a large knife. I also have another mental picture of the same pair of black-leather lovebirds. Later that same day I was nearly run over as the big fellow, with the young blonde hanging affectionately onto his back, raced his motorcycle right up the main pedestrian walkway on the Montana State University campus. Here is the blank part of the tape: Although I have a clear memory of his ominous size and black clothing, I have no recollection whatsoever of his face. In my memory he is in the same faceless category as Darth Vader. Nor do I remember the color or model of the motorcycle that sped up the pedestrian pathway, though I'm betting it was a Harley.

Besides the many blank spaces in my memory, there are some places where what is filled in cannot be true. I realize this every time I have a conversation with my ex-wife about something that happened when we were both in our twenties: Her version is often completely different from mine, and sometimes she has actual evidence to suggest that the version seared into my memory has been edited, usually in a way that makes me look better.

What We Remember Depends on Who We Are at the Moment

In this chapter, I am going to talk about some research on how the human mind crunches information about our experiences with other people. On any given day, you may pass hundreds or thousands of people on the streets, in restaurants, supermarkets, churches, athletic clubs, or public restrooms—or on the pedestrian pathways of a college campus. Some of those people talk to you, overloading your mental capacities with still more information—about who said what to whom, when, where, and why. As they talk, these various people adopt different postures and diverse tones of voice. Sometimes they are telling you facts, sometimes they are kidding, and sometimes you think they may be lying. Some of them are wearing flashy colors, some are dressed conservatively, some have on black leather jackets, and some have messages written on their T-shirts. Some bear the aroma of a pleasant perfume; some reek of alcohol, tobacco, or motorcycle grease.

Most of the daily zillion bits of social information go in one ear (or eye or nostril) and out the other. But some people, and some of the things they say or do, sear themselves deeply into our memories, so deeply that we remember the specific details of certain social interactions decades later.

Cognitive Science Meets Evolutionary Psychology

Why do we remember some people and some social situations and forget others? And why are some of our memories hopelessly distorted? Cognitive psychology is the branch of the field that deals with these questions. The field of cognitive psychology developed hand in hand with the field of computer science. In fact, computers were developed by researchers at the interface of psychology, philosophy, and mathematics—folks who wanted to build machines that could think like humans. So it's no surprise that when psychologists think about how the mind works, we often use computers as a metaphor, envisioning our brains as information-processing machines.

One way to think about mental information processing is to imagine a series of progressively finer-meshed filters. At the first stage, an array of sensory mechanisms (for detecting temperature, volume, or color) feed a broad stream of information into the brain's attentional filters. Our brains choose only a small portion of that information for conscious attention. That information is "encoded" or categorized (for example, Is this particular combination of sounds and shapes a dog barking, or a man yelling?). At the next level, only a small percentage of the information we encode makes the cut to get stored in our long-term memory banks. If that information is extremely important, such as the information that someone is attracted to us or that someone we know has died, it will get privileged processing, and may even become one of our flashbulb memories. But many of the things that make it past the first filter do not make it into long-term memory (the names of most of the people we are introduced to at a party, for example).

Cognitive psychologists began with the assumption that information processing is information processing is information processing. On this view, categorizing a string of letters as "cat" and not "cot" involves the same mental processes as categorizing an arrangement of facial features as anger and not happiness. There is parsimony in assuming that the brain processes all information in more or less the same way and that those processes are more or less the same as those used by a computer. But as I discussed in Chapter 6, a theory can be too parsimonious, and research on domain-specific processing now suggests that the brain crunches different kinds of information in qualitatively different ways.

The philosopher David Hume famously said that reason is the "slave of the passions." Given what we now know about modularity, I would modify that slightly. We have not just one central reasoner inside our heads but several. Which details we notice and remember and which ones we distort depends on what is most functionally relevant to the subself currently in control. Although we have only one motivational subself in the cockpit of consciousness at any moment, the others have their radar systems running in the background. When the swinging single subself is at the controls, we may be thinking about an attractive person who just passed on the street and remembering a pleasant date with someone who looked like his or her cousin. But if a group of scary-looking teenage hoodlums with angry smirks walk onto the scene, the night watchman takes the pilot's seat, and we may begin looking for ways to avoid crossing their path and remembering the police officer who passed a few steps back.

The Evolved Computer Inside Your Head

Imagine you are on the subway with a man sitting across from you. You may not notice whether he is wearing a plaid jacket, but if he is making an angry face in your direction, a flashing red light will go off in your brain in less than a second.

What the mind chooses to prioritize is a question at the interface of cognitive science and evolutionary psychology. Along with my colleagues Vaughn Becker, Steve Neuberg, and Mark Schaller and a team of crack graduate students and former students, I have been doing a lot of research at the interface of these two fields.

In fact, I have already talked about some of this research. As we saw in Chapter 4, our own motivations, such as fear or amorous emotions, can lead us to project completely different meanings onto identical facial expressions. Thus, frightened people saw anger and amorous people saw sexual receptivity in the same neutral faces. And in Chapter 2 we saw that women—although they will spend time looking at good-looking men—do not remember them very well later, whereas men looking at attractive women do remember them. We think that sex difference in memory is linked to men's and women's different mating strategies: A man can in theory reap more reproductive profits and pay lower costs from having a relationship with an attractive stranger; a woman needs to make an informed decision. Perhaps staring at a handsome stranger increases the odds he will introduce himself. But if he does not, a woman is not likely to chase after him or even to waste cognitive resources thinking about him.

The essential point—which will be borne out as we look at more of the work that my colleagues and I have done—is that in order to understand how and what the human mind computes, one must place it in an evolutionary, ecological context. If we want to know why the mind works in a certain way, we must ask how and in what circumstances it would be beneficial to do so. Our brains seem to allocate resources in ways designed to best promote survival and reproduction.

In the rest of the chapter, I will describe a few of our interesting findings.

In one experiment, we gave our subjects a very easy task: They had to look at a face on a computer screen and press the "A" key if the person shown was angry and hit the "H" key if that person was happy. The task was especially easy because all the faces were wearing very clear emotional expressions (either contorted with anger or wide-eyed and smiling). And our subjects were very good at it, hitting the correct key almost every time, and doing so in less than a second. But some of the decisions were even easier than others. If the face on the screen was a man, people almost never made a mistake when he was angry. But when he was smiling, they made mistakes almost 10 percent of the time, even as they took longer to make a choice. If the face on the screen was a woman, on the other hand, the results were reversed: People were faster and more accurate in recognizing happiness on a woman's face.

Was it because men are better at expressing anger whereas women are better at expressing happiness? To address this question, Vaughn Becker made up some artificial people, using a program called Poser that creates lifelike human faces. The program allows the user to choose the extent to which a face is sex-typed, so Vaughn could make faces that ranged from completely androgynous to highly masculine or highly feminine. Vaughn was also able to give the face any degree of emotional expression. Even for these computer-generated faces, our subjects' judgments were still quicker and more accurate in recognizing anger on a male face and happiness on a female face. Vaughn also found something else fascinating: If he took a completely androgynous face and made it look just a little bit angry, people overwhelmingly judged it as a man. If he took the same androgynous face and made it look slightly happy, people judged it to be a woman.

We explained the results, especially the rapid detection of angry men, in evolutionary terms: Emotion researchers since Darwin have thought about emotional expressions as a coevolved pairing of mechanisms to transmit and receive emotional cues. An angry expression communicates threat and can be a warning that helps two people avoid a direct bloody confrontation, but it only works if the receiver's brain recognizes it. Why would it be functional to recognize a man's anger more rapidly than a woman's? Because men are more potentially violent, so we do not want to miss an angry expression on a man's face. Why would it be adaptive for men to communicate their anger accurately? Men are more likely to be involved in conflict with one another, and their ability to communicate anger could influence their position in the dominance hierarchy (which for men more than for women can translate into reproductive success).

When we submitted the results for publication, a couple of the reviewers were unconvinced, arguing that learned sex roles could explain the effect. We needed to tease apart the competing coevolutionary and sex-role explanations, which can be tough, because the two kinds of influence often work hand in hand. In this case, however, computer software came in quite handy for making the distinction. We again used the Poser program to generate completely androgynous faces. But now we took those sexless faces and put them on bodies bearing sex-role accoutrements. The faces were identical, but the male body had a thicker neck and broader shoulders and was adorned in a suit and tie; the female body was adorned in a dress. In another variation, we did not add any cultural cues, but simply tweaked the facial features to make them more masculine (thicker eyebrow ridge and larger jaw) or feminine (rounder brow ridge and smaller jaw).

When we put the androgynous face on a male body wearing a suit and tie, it was in fact judged to be a man. If the sex-role explanation were correct, then, and there is a learned mental association between male and anger, the person judged to be a man should also have been judged as more angry. But he was not. When we gave the androgynous face a man's eyebrow ridge, on the other hand, the face was judged as only slightly more masculine but much angrier than the purely androgynous face. This pattern of results supports the idea that what makes a man's face look more angry is its morphological structure—that naturally prominent brow—not the fact that people simply associate masculinity with anger. In other words, men's faces are constructed to better communicate anger, and our brains are constructed to quickly spot anger in a man.

An angry face grabs your attention because it is particularly relevant to your night watchman subself—you do not want to be caught sleeping if there is a potential threat to your life and limb. Other subselves are attentive to different social signals. The swinging single subself is tuned to attractive members of the opposite sex. But these subselves are not equally strong in all people. Some people are happily monogamous, for example, and do not spend much time scanning for new mating opportunities. So in other research we conducted with Lesley Duncan and Mark Schaller's team, we decided to measure how sexual pursuit interacted with attention. For this research, we used what is called the "change detection" method. Subjects in this experiment looked at eight faces on the screen and we asked them to press a computer key whenever one of them changed. Sometimes an eye or a nose on one face would repeatedly disappear and reappear on the screen. Those changes happened fast, with the normal and distorted versions flipping back and forth several times a second. A disappearing nose would seem easy to spot, but it can be surprisingly invisible unless you are paying attention to the particular face as it changes. We found that people were more likely to notice a change in a physically attractive member of the opposite sex, but there were two important qualifications. First, this selective attention effect was only found among men. Second, the effect was found specifically among men who adopted a nonmonogamous, "unrestricted" approach to dating. These are the men whose swinging single subself never goes to sleep.

Remembrances of Things Not Past; or, Regrets, I've Had a Few

Reminiscence can take two forms. Sometimes when we reminisce, we think about things that actually happened, like that three-way kiss or the painful call from my mother. But sometimes we think instead about alternative realities, about what it might have been like if we had walked up a different life path at some time in the past. Indeed, sometimes we remember more about the imagined alternative than we do about events that actually happened.

As a younger man, I had a passionate crush on a beautiful but reticent young woman. One afternoon, though, she seemed ready to overcome her reluctance and started leading me back to her bedroom. But as I thought about my current girlfriend, I decided to pass up the opportunity. Nevertheless, I still wonder what would have happened if I had stayed. Would it have turned into a pleasant memory of a short and passionate affair, or would it have completely disrupted my life?

Many of my male friends have a counterfactual reminiscence very much like that one—of a time they almost got to sleep with a woman they found highly attractive. I wondered whether women have similar reminiscences.

On a sabbatical many years later, I had a conversation with Neal Roese, who was then a professor at Simon Fraser University in British Columbia. Neal is probably the world's foremost expert on counterfactual reminiscences—musings people have about choices they did not make and about how things might have been if they had chosen differently. He and I had independently started wondering about possible sex differences in these counterfactual thoughts.

Previous researchers had not found many interesting sex differences in counterfactual thinking, but as Neal and I talked it became clear that researchers might have been asking the question the wrong way. Earlier studies had focused on people's counterfactual thoughts about possible achievements ("If only I had studied harder or made that home run," for example) and had not probed much into counterfactual musings about relationships. Working with Neal's colleagues Ginger Pennington, Jill Coleman, and Maria Janicki and with my colleague Norm Li, we took a more focused look at people's counterfactual thinking in the romantic domain. Across several studies, we asked participants to think about either their past romantic relationships, their academic achievements, their friendships, or their relationships with their parents. As they thought about these different relationships, we asked them to consider the question, "Is there something you wish you had done differently?"

When it came to their parents and to their school careers, both men and women had about twice as many regrets over inactions, things they should have done but did not, than over actions, things they actually did but wished they had not. When it came to romantic relationships, though, men and women were very different. Women were much more likely than men to have regrets over things they had done (getting involved with that self-centered bastard despite Mom's warnings, for example). The vast majority of men's regrets over relationships, on the other hand, were about actions they did not take (a time they did *not* get more intimately involved with some desired damsel).

From an evolutionary perspective, this makes a lot of sense: Men incline a bit more toward promiscuity, and women to careful choice. When women make a bad romantic choice, they remember. And perhaps by remembering, they will do a better job of avoiding those mistakes in the future.

A Conceptual GPS Reading: Where Are We Now?

In the first chapter, I promised a journey from the gutter to the stars. After that, we had a whirlwind tour of the gutter, stopping for scenic overlooks down on sex, murder, and racial prejudice. We saw how gaping over *Playboy* centerfolds, though based in very natural inclinations, could wreak havoc on the landscape of our relationships in the modern media-heavy world. We drove into the darker mental neighborhood of homicidal fantasies, getting a few glimpses of natural inclinations that send some of our neighbors on unwanted side trips to emergency rooms and prison cells. And then we descended deeper into the jungle of human nature, taking a gander at the different areas of prejudice.

In Chapters 6, 7, and 8, we have shifted our gaze up into our own minds. We have seen that what is in there is not a blank slate and not a blueprint but a coloring book, with some guidelines drawn before birth and some spaces awaiting the artistic inputs of life experience. And we have seen that there is not just one executive self inside our heads but sets of sometimes segregated subselves, each designed to perform different essential tasks, such as watching out for bad guys from other neighborhoods, getting along with our neighbors, and taking care of our families. Those different subselves come on line at different phases of our life's journey, and they take turns in the driver's seat, depending on the threats and opportunities in the current landscape.

In the next few chapters, we will move away from the individual's perspective and take an increasingly aerial view. We will now begin looking at how those simple selfish biases inside our heads connect us to other people, influencing conformity to social norms, socially motivated consumer decisions, and even our decisions to go to church. In the last chapters, we will move still higher, gazing down to see how these earthbound local motivations interconnect with economics and the emergence of societal order.

Chapter 9

PEACOCKS, PORSCHES, AND PABLO PICASSO

As a four-year-old, my younger son couldn't have cared a whit about what we paid for his toys. New or used? No matter. Free? Fine. One day, he and I cut up an old cardboard box and some construction paper to make a rickety tunnel for his toy cars. Cost of materials: maybe twelve cents. He couldn't have been more euphoric if I'd dropped a couple of hundred bucks on the latest fancy kid's gizmo at FAO Schwartz. But I know his obliviousness to market value won't last. When I was about his age, I would happily accompany my mother into a place called John's Bargain Store, hoping she might find some goodie for me. By the time I was eight, though, I adamantly refused to even enter this lowly discount emporium, and I can vividly remember one tantrum that no doubt humiliated her right in the middle of the Steinway Street shopping crowd. I'm sure Mom was imagining the onlookers thinking, "Oh, my goodness, Gracie, Irene's forcing her poor little boy to shop in that dumpy little thrift store!"

Materialistic karma came back to haunt me when my older son was approaching his teens and would throw a similar tantrum at the mere suggestion that he might choose a pair of athletic shoes costing less than $120 (and that was back in the 1980s, when my income was a lot lower and a dollar was worth a lot more than it is today). The classic parental lament was that in a few months, the same shoes could be bought on sale for $20. But when I tried to steer my son to last season's model, he would point out that they were no longer "cool." Michael Jordan, and all of America's youth, had moved on to a slightly different model whose apparent coolness was lost on parents but blatantly obvious to young boys.

Bigger and Costlier Things

Of course, adults become more economically rational, don't they? My friend Ed Sadalla, the professor who first showed me a copy of Wilson's *Sociobiology* in 1975, told me a story about a friend of his who lives in Pacific Palisades, a swell Los Angeles neighborhood flanked by Malibu and Santa Monica. Ed's buddy was in the market for a new set of wheels and wanted Ed's opinion about a particular model of Lexus. Ed is a good person to ask such a question. He has driven some classy automobiles in his time, including a Porsche and a classic Volvo sports car. Ed opined that the Lexus in question was in fact a very well made vehicle, but he suggested his friend consider another car that was virtually the same machine but would save his pal several thousand dollars. The only real difference was that the alternative would bear the brand name "Toyota" instead of "Lexus." According to Ed, his friend wrinkled his nose slightly and said, "Oh no, I could never do that!"

Was Ed's friend just being taken for a ride by the Lexus label? If we calculate the value of both cars from the perspective of *Consumer Reports*, it is clearly less rational to buy the Lexus. Of all the pertinent features—acceleration, braking, comfort and convenience, fuel economy, reliability, interior comfort, and safety features—the Toyota Camry scored lower on only one: the sticker price. And given that mechanics charge more to work on luxury cars, the Lexus would have another disadvantage: It would cost more to maintain in the future.

But economic utility involves more than just value for a dollar. Economists define utility as expected future satisfaction. As Thorstein Veblen noted over a century ago, some people gain a great deal of satisfaction from displaying their wealth. Indeed, if you want to advertise your affluence, a certain amount of wasteful frivolity in your spending patterns is quite useful—it indicates you have money to burn. *Consumer Reports* has a category, "Best Buy," that we usually think of as a good thing. But from the perspective of Veblen and of Sadalla's friend in Pacific Palisades, economic frugality could, ironically, reduce a car's value—counting as a liability for those who want to conspicuously burn their money.

In fact, compared to many other automobiles, the Lexus hardly registers on the wasteful consumption scale. Consider the Porsche Carrera GT. That particular vehicle has very little cargo capacity, only two seats, and a ten-cylinder engine that gets terrible gas mileage and is frightfully expensive to repair. The manufacturer's list price is $440,000, not counting the $14,800 dealer prep charge and the $5,000 delivery charge. And that seems modest compared to the $643,330 sticker price on the twelve-cylinder Enzo Ferrari.

Is conspicuous consumption a symptom of a decadent and overly materialist capitalist society, as critics often claim? Probably not. In his classic *Theory of the Leisure Class*, Veblen documented instances of conspicuous consumption from Iceland to Japan. On the Trobriand Islands, the chiefs give away valuable jewelry to demonstrate their greatness. On New Guinea, the big men compete with one another to give away the largest number of their highly valued pigs. Among traditional native tribes in the Pacific Northwest, such as the Kwakiutl, the headman occasionally throws a potlatch ceremony, in which he systematically gives his most valued possessions to other people and may even end the ceremony by burning down his own house. And history is rich with even more notorious examples of conspicuous consumption, as preserved in the golden thrones, elaborate artworks, and giant pyramids of the Egyptian pharaohs, the immense palaces and ornate gold jewelry of the Incan potentates, and the extravagant and ostentatious palaces of the Indian maharajas.

Modern-day researchers in the field of consumer behavior often point their fingers at the big bad guy of twentieth-century social science—American culture—as the alleged perpetrator of conspicuous consumption. But as a small minority of consumer researchers are beginning to point out, conspicuous consumption is yet another human behavior whose roots can be found only by looking more deeply, examining not only the parallels between modern Americans and people in other societies but also our links with animals who have never turned on a TV or watched a movie.

Peacocks and Sexual Selection

In developing his theory of evolution by natural selection, Darwin was troubled by a different kind of conspicuous display—the beautiful and attention-grabbing plumage on birds such as peacocks. The peacock's tail is enigmatic in at least two ways. First, the bird has to invest a lot of energy to build his ostentatious tail, meaning he has to spend a lot of extra time searching for food. Second, the bird, which is native to India and Sri Lanka, is hunted by tigers, cheetahs, lions, hyenas, and wild dogs, all eager to turn a large and tasty bird into dinner. And then there is the dread *Homo sapiens*, who hunts down the remaining wild peacocks for their tasty meat and colorful feathers. To all his predators, the peacock's display is analogous to wearing a neon sign that says, "Eat here!"

If evolutionary forces favored features that promoted survival and reproduction, how could any animal evolve a characteristic that increases its odds of dying young? An important clue comes from a statement by a wildlife official from the Indian Punjab, who said, "It is the easiest to kill a male peacock during the mating season, when it dances around in the open and can be easily targeted." So the peacock shows off his brilliant display only during the mating season. And if you have ever seen a mixed-sex crowd of peafowl, you know that the brilliant display is only found on one sex—the male. Peahens are colored to better blend with their surroundings and have a tail built for speed rather than show. So the male is showing off to attract mates.

Peacocks are uniquely beautiful, but the species is hardly alone in having show-offy males. That mockingbird singing outside your window all night long (possibly attracting the attention of hungry cats) is a male. The bighorn sheep in Rocky Mountain National Park hurling itself headfirst at full speed against another bighorn—male. The Siamese fighting fish flaunting its colorful fins and ever ready for a brutal altercation—also a male. It is not that females never show off, for reasons I will discuss below, but showing off is like homicide: There is a small chance that the perpetrator is the maid, but any aspiring Sherlock Holmes would fingerprint the butler first.

To explain how traits like the peacock's tail could evolve, Darwin invoked the concept of sexual selection, which we have already encountered in our discussion of aggression in Chapter 3. Although evolution is often mistakenly dubbed "survival of the fittest," the name of the game in evolution is not survival but reproduction. To be fit in an evolutionary sense, an animal needs to survive long enough to reproduce. But survival without reproduction does not enhance fitness. A celibate life, no matter how long, does not contribute genes to future generations. On the other hand, an animal that lives fast, attracts mates, and has offspring is an evolutionary success story even if it dies young.

Sexual selection comes in two flavors. Some traits (such as a peacock's tail) enhance an animal's fitness by attracting members of the opposite sex. Other traits (like antlers or horns) may enhance fitness via an indirect route: by helping the animal compete with members of its own sex.

Why is it that males have these showy and costly accoutrements more often than females? The answer links back to another concept I discussed in Chapter 3, differential parental investment. Recall that whenever one animal makes a relatively higher investment in the offspring, that animal tends to be relatively choosy about mating. Females have an intrinsically high investment because they produce eggs and, in the case of mammals, carry the growing young inside their bodies and later nurse them. Males can, in theory and often in practice, make a very small direct investment in the offspring—the energy it takes to produce sperm. Hence, females are more likely to be choosy about mating, and males are more likely to show off.

Dominance and Sexual Attraction in Humans

In the first chapter, I briefly mentioned the first research Ed Sadalla and I conducted with Beth Vershure to test the applicability of sexual selection to human beings. Although biologists since Darwin's day had collected numerous examples of sex differences in dominance and had linked the difference to female preference for dominant males, no one had ever asked whether the same phenomenon applied to human beings. So we conducted an experiment designed to examine this possibility.

We brought college students into the lab and showed them videotapes of two different members of the opposite sex. One of the people they saw moved in a manner designed to convey submissiveness. The other used the nonverbal signals of social dominance. For example, a woman would watch a video in which a man enters an expansive office in which another man is seated behind a large desk. In the submissive condition, the man entering the office sat near the door, far away from the man behind the big desk. During the conversation that

ensued, the visitor held his body in a rigid position with his head par-
tially bowed, clutching a sheaf of papers and frequently looking down
at the floor. In the high-dominance condition, a different man would
strut into the same office, pull his chair right up to the desk, and move
his hands and body in a relaxed manner, all the time leaning in close
to the man behind the desk. Male subjects saw an identical pair of
videos, except that the actors were women.

We told subjects that we were testing their abilities as amateur personality psychologists, their aptitude for judging someone's personality with only a small amount of information to go on. We told them
that each person entering the office had been given an extensive series
of psychological interviews and tests and that their job was to guess
what the person was "really like," based on observing the person's behavior in a silent video. So each subject rated both the dominant- and
submissive-acting targets on a number of personality traits, such as
"dominant vs. nondominant," "feminine vs. masculine," and "warm vs.
cold." We were in fact most interested in what they had to say about
how sexually attractive each person was.

The results were clear for women judging men: When the target
acted dominant rather than submissive, our female judges rated him
as more sexually attractive and desirable. How did men react when a
woman acted dominant? According to one theory popular at the time,
which claimed that women avoid acting dominant because they fear
it will make them unattractive and masculine, acting dominant should
have hurt. But whether the woman was dominant or submissive actually made no difference to her attractiveness. In three other studies,
we used different methods to make the target appear dominant (being
a tennis player who dominated opponents, or being rated by a team of
psychologists as powerful, commanding, and masterful). In each case,
we replicated the original pattern. Dominance never had an effect on
a woman's attractiveness, but it reliably made a man significantly sexier to women.

Since that time, other researchers have found similar effects for other forms of social dominance. For example, John Marshall Townsend and Gary Levy found that high-status clothing boosted a man's attractiveness, so that an unattractive man wearing a nice suit and a Rolex watch was more desirable to women than a handsomer guy dressed as a Burger King employee. Men judging women, on the other hand, preferred the good-looking woman, regardless of whether she was dressed up to impress or dressed down to serve French fries.

Filling out the story, anthropologists have collected data from other societies and other time periods showing that around the world and throughout history, high-status men have had access to more wives (in polygynous societies) and more attractive wives and extramarital partners (in officially monogamous societies).

Flashing the Cash

A male can stand out by being physically dominant over other males, like an alpha chimpanzee, or he can demonstrate his superior qualities to females in other ways, peacock style.

Jill Sundie, now a marketing professor at the University of Texas, was a graduate student of mine who discovered evolutionary psychology after studying economics. Jill suspected that the pursuit of money—and the visible spending of it—has something in common with a peacock growing an elaborate tail. We explored those commonalities in a series of studies with Vlad Griskevicius, Josh Tybur, Kathleen Vohs, and Dan Beal.

In one study, we asked students at two different universities to think of a time when they had witnessed someone engaging in conspicuous consumption. The majority of students thought of a man—buying a flashy car or picking up an oversized group tab at a restaurant, for example. Is this because men simply have more money to consume things in general, as one of Jill's reviewers confidently suggested when she submitted the findings to a marketing journal? Not quite. When we later asked a similar group of people to think of the person they knew who most liked to shop, the majority nominated a woman. So people perceive women as liking to spend money, but men as liking to throw it around in conspicuous ways.

Following the peacock analogy, we suspected that men's conspicuous consumption was a form of showing off linked to mating. To examine this possibility, we ran several other experiments. In one, we asked participants to imagine they had just received an unexpected windfall of $5,000. How much of the money would they spend on purchases that might convey their newfound wealth, such as a new watch, a new cell phone, or a nice vacation to Europe? Before asking them how they would spend the money, we put some of the subjects in a mating frame of mind by asking them to imagine an ideal first date with someone they found highly attractive. Others looked at photos of buildings.

Romantic motivation had a different effect on men than on women. Men in a romantic frame of mind chose to spend more of their newfound $5,000 on conspicuous purchases (but not to spend more on inconspicuous purchases, such as tissues, cold medicines, or kitchen supplies). Women's spending was unaffected by feelings of romance.

It is not that women are oblivious to the influences of romantic motivation. In another series of studies with Bob Cialdini and Geoffrey Miller, we asked men and women not about spending but about volunteering to help out with a local charity (working at a homeless shelter, for example, or teaching underprivileged children to read). Romantic motivation did not affect men's charitable inclinations, but it did boost women's tendency to want to help others. In several other studies, we replicated the same finding: Romance made women more inclined to help other people out. The only time romantic motives led men to act more altruistically was when the help could make them look like a hero (jumping into icy water to save a stranger who had fallen from a boat in a storm, for example, or distracting a grizzly bear who is attacking a stranger).

Creative Genius: How Is Picasso Like a Peacock?

Pablo Picasso produced 147,800 works of art in his life, more than any other artist in recorded history. And he did not just knock off the same painting over and over; he kept recreating himself, with different styles in his Blue Period, his Rose Period, his Cubist Period, and his Surrealist Period. When art historians looked at Picasso's life, they noticed something else going on: Each of his new creative periods was accompanied by a new mistress. His latest paramour was always a younger woman who, like the beautiful Dora Marr, served as an inspiring muse for Picasso's new style. And Picasso was not unique in this regard; historian Francine Prose observed that many historically creative figures, including Salvador Dalí, Friedrich Nietzsche, and Dante Alighieri, were also inspired by muses of their own.

Evolutionary theorists have offered several possible hypotheses about the origins of creativity. Most have presumed that creative intelligence assisted in survival (for example, a creative mind could create a novel way to catch a fish or shake fruit out of a tree). Steven Pinker suggested instead that creative abilities are simply by-products of other cognitive capacities. But in his book *The Mating Mind*, Geoffrey Miller argued that these hypotheses failed to explain the sheer amount of time and effort people devote to creative exploits that produce no tangible benefits (such as writing poems, composing music, or painting pictures), meanwhile ignoring tasks that could directly produce food or other immediately useful goodies.

Along with Bob Cialdini and I, Vlad Griskevicius set out to reproduce the "muse effect" in the laboratory. We reckoned that creativity in human males, like feather displays in male birds, might be linked to sexual selection and might be triggered by reproductive motivation. In birds, sexual motivation is limited to the springtime mating season, but in humans, mating urges are less seasonal and can be triggered by awareness of an available and attractive member of the opposite sex.

In one experiment, we brought college students into the laboratory and asked them to write a short story about an ambiguous picture—such as a colorful abstract painting or a cartoon drawing of two men chatting at an outdoor café. Before writing their story, half of them were put into a mating frame of mind by viewing six photos of highly attractive members of the opposite sex and then choosing the one they would most desire as a romantic partner. After they made their choice, we left the photo of their selected romantic partner on the screen and told them to imagine what they would do on an ideal first date with this person. The other half of the subjects were in the control condition; they just saw a photo of a street with several buildings and were asked to write about the most pleasant weather conditions for walking around and looking at the buildings.

The students' stories varied greatly in their creativity. For example, for the café cartoon, one guy said, "These two people work together and are on a break from work at a coffee shop."

Another more creative type said:

Nigel is trying to decide whether or not to get a nose job. He just can't decide. However, his friend Reginald had one and his nose was simply stunning. Reginald is a very particular sort of fellow you know. That latte he's drinking had to be just so. Soy milk with a dollop of foam and merely a whisper of cinnamon. Too much of one ingredient might completely throw off the balance of his day. When one is so particular about cinnamon, you could only imagine how he'd prefer his nose. All of these things Nigel noted to himself as Reginald went on and on.

In describing one of the abstract paintings, another student wrote:

The setting is a seedy, underground jazz club, where bands have to compete with drug dealers for the patrons' attention. A good quintet is performing, with a tenor saxophone, two trumpets, a trombone and a drummer. The instruments are old and worn, but the music that they make is enough to turn the attention of the crack dealers and the junkies. The music is haphazard and at times seems arrhythmic and amelodic, but it fits the scene like a velvet glove.

After we collected the stories, we showed them to other students, who rated each story on the extent to which it was creative, original, clever, imaginative, captivating, funny, entertaining, and charming. When we analyzed the results, we found that mating motivation had absolutely no effect on women's creativity. But the mating prime really got the men's creative juices flowing. Although the men in the control condition were slightly less creative than the women, the guys thinking about mating were positively inspired. It was not that the mating-motivated men wrote more, but that what they wrote was judged more clever, imaginative, and original. In another study, we found that men thinking about mates also racked up higher scores on standard tests of creativity, such as finding remote associations to words or devising new uses for familiar objects.

So these studies established that temporary activation of a mating motive can have the same effect on humans as the mating season has on peafowl; in both cases, mating opportunities inspire males to strut their stuff.

He's a Rebel, but in a Good Way

One of the characteristics of great artists like Picasso is that they are constantly trying to break with tradition, striving not merely to paint or write well but to paint or write in some radically new way. Rivera and Neruda broke traditions not only in their art and poetry but also in their politics. As it turns out, rebellious thinkers are often quite attractive to women, even when they are not artistically inclined, as was the case with Ernesto "Che" Guevara.

Of course, the key is not simply to say things that sound crazy but to offer new ideas, preferably ones that will improve society. After growing up in a wealthy family and completing medical school, Che Guevara took a trip around South America. He was so offended at all the poverty and oppression he saw that he decided to dedicate himself to fighting injustice and promoting equality (whether or not one agrees with his politics of communist revolution or with his behavior after he came to power in Cuba, it is important to appreciate that Che did not start out looking for trouble and change for its own sake; he had a higher goal).

Chad Mortensen and Noah Goldstein joined us to study the effects of romantic motivations on nonconformity. We reckoned that nonconformity is another tactic that men can use to stand out from the crowd and attract mates. But as in the case of political rebellion, we reasoned that mating motivation would not simply inspire men to disagree with others in random or senseless ways. Instead, we expected it would lead them to disagree with others in ways that would make them appear to be uniquely discriminating—to manifest the qualities of a good leader rather than those of either a docile follower or an angry rebel without a cause.

To measure people's inclinations to stand against or go along with group opinions, we created an experimental situation reminiscent of Solomon Asch's classic line-judging study, in which a subject had to give his judgment on the comparative lengths of lines after hearing five other people say that a longer line was actually shorter. In our study, subjects were asked to judge how interesting they found an artistic image. But before they could register their opinion, they first got to see the judgments of several other members of their group (who tended to agree with one another that the image was either rather uninteresting or quite interesting). Did the subject go along with the group judgment or stand alone? The answer depended on whether said subject was a he or a she, and also on his or her motivational state.

Some of the subjects were feeling fearful, after having imagined a scenario in which they were alone in a house late at night. If you were in this condition, you would imagine lying in bed in the dark and recalling a news story about several unsolved murders in your neighborhood, then hearing strange noises outside, followed by sounds of someone entering the house. After calling out and receiving no reply, you try to use your phone and realize the line is dead. As the story ends, you hear strange cackling laughter, and then see the shadow of an intruder entering your bedroom.

Other participants were put into a mating motivation. If you were in this condition, you would imagine being on vacation and meeting the person of your dreams. After spending a romantic day together, during which you become increasingly infatuated with one another, the scenario ends as you kiss passionately on a moonlit beach.

If you were in the control condition, you would imagine finding some lost concert tickets just in time to go to a show with a friend.

As we had predicted, fear led both men and women to conform more to the group's opinion. (This result fits with numerous findings that animals and humans pull together under threat.) But romantic motives had different effects on men and women. Romantic mood, like fear, boosted women's conformity. But for men, a romantic mood inspired them to stand against the group's opinion. Furthermore, they did so in a very strategic way. Romantically motivated men only went against the group opinion when doing so could make them look good, by expressing a uniquely positive opinion when the group had been negative (Ah, but I disagree; I can see the beauty in that). And in a later study, we found that romance inspired men to go against a group only on subjective judgments, for which there is no objective right or wrong answer (Do you prefer paintings by Vincent van Gogh or Claude Monet?). When there was a clearly correct answer, and they could be proved wrong (Is it more expensive to live in New York or San Francisco?), men went along with the consensus.

So again, we see that mating motives inspire men to show off in another way: By selectively demonstrating their independence from group opinion, particularly when they can do so in ways that make them look uniquely good rather than oddly different.

The Many Faces of Showing Off

So men respond to romantic motives by showing off: by spending money in more conspicuous ways, by acting like heroes, by standing firm against group opinion in ways that could make them look good, and by strutting their artistic creativity. What is important is not simply that men are show-offs but that they flash their feathers in especially ostentatious ways when they are in a mating frame of mind. Is it just a coincidence that human male showiness responds to mating motives in the way that males in other species respond to the mating season? Perhaps, but it seems unlikely. Remember, women prefer to mate with men who stand out from the crowd. Thus the payoff is the same for male humans as it is for peacocks and bighorn sheep, and it is the payoff that drives natural selection at all levels—an increase in attractiveness to the more discriminating sex, who will be disinclined to mate with a male who does not demonstrate his worth. And for men, as for males of other species, showing off is expensive. Conspicuous consumption is literally paying money to get noticed. Dedicating years to becoming a creative artist, as Picasso and Rivera did, can involve periods without food on the table. And it is often the tallest grass that gets cut first. Standing against group opinion, even in noble and positive ways, can result in harassment, imprisonment, and even death. (Martin Luther King Jr., like Che Guevara, lived fast and died young, but both, like Picasso and Rivera, were highly attractive to women.)

What About Female Displays?

Why don't women respond to mating motives in the same way that men do? In several studies, we tried to find a condition in which a woman would be more creative after thinking about romance. It only worked in one special case: when the women were thinking about a long-term relationship with a man whom they had been dating for a while and who had impressed their friends and relatives as a good catch. Then and only then would a woman show off her creative side. In this regard, consider one example of a female being inspired to creative brilliance by romance. During her twenties, Elizabeth Barrett Browning had not yet hit her stride as a poet. But then she received a fan letter from Robert Browning, who professed to be in love with her. It took a year, and many more letters, for Robert to convince Elizabeth of the sincerity of his intentions, at which point she agreed to elope with him. It was at this time that Elizabeth was inspired to write *Sonnets from the Portuguese*, regarded by critics as her best creative work.

Because human males often invest in their offspring, they also exercise choice in picking females as mates, at least for long-term relationships. But even then, men and women choose one another based on different criteria, as I discussed in Chapter 2. Women contribute more directly to caring for the offspring, donating their own bodily resources, and hence many of women's flashy purchases are designed to make them appear physically attractive and healthy. And as I mentioned in describing our conspicuous consumption studies, women did respond to thoughts about mating by showing off their nurturant qualities—increasing their desire to take care of others. By display-ing their nurturance, women demonstrate a trait likely to be valued in a mother.

Peacocks, Porsches, and the Meaning of Life

The links among such broadly different forms of display—between birds strutting their brilliant feathers and humans displaying artistic genius, nonconformity, heroism, and conspicuous cash-flashing—highlight the profound interconnectedness of nature. They show how broad abstract principles derived by biologists studying animals in the jungle weave together with economics, politics, and even poetry. Freud was onto something when he attributed mankind's great accomplish-ments to the sublimation of sexual urges, but because he was working without these new broad principles, he missed the central point. As we emphasized in Chapter 8, reproduction ain't just about having sex. The noble and brilliantly creative things human beings accomplish are not simply sex drives gone astray; they are elaborate forms of fore-play, intimately tied to the processes by which our ancestors chose which genes were going to make it into future generations.

These findings also highlight that there are many roads to repro-ductive success, not all best traveled in a Porsche Carrera. If your kid wants to become an artist, it could ultimately make him more attrac-tive than buying an expensive car when he finishes medical school. And you certainly should not think of creative play as mere distrac-tion; we are wired up to deeply appreciate it and to respond favorably to those who are especially good at it.

There is another reason why you should not feel bad if your son is not driving a Porsche Carrera. Very rich and flashy guys do in fact get more action than guys in Hyundais, but studies by Jill Sundie and her colleagues suggest that people associate extreme extravagance with short-term mating strategies. It is those men who are inclined to be playboy types rather than the stay-at-home dads who are most motivated to show off their extravagance to the opposite sex. And women understand this. They view men who buy flashy expensive cars as one-night-stand types. In some related research, we explored the relationship between income and desirability as a marriage partner. We found that most of the difference in desirability shows up between poverty and a lower-middle-class income. Once a guy brings in a middle-class income, he is almost as attractive a marriage prospect as a very wealthy man, and he is probably more likely to come home at night and help take care of the kids.

Climbing the Pyramid

In laying out the map of what was to come, I promised that looking in the gutter, at base topics like sex and aggression, and at the links between humans and other savage beasts, would help us understand the so-called higher reaches of human nature. In this chapter, we have seen that some of those higher aspirations—the creativity and independence of judgment that defined Maslow's self-actualized icons—are intimately connected to the same evolutionary mechanisms that underlie sex and aggression. In the next chapter, we'll consider a very different connection of this sort, exploring a link that my Catholic school nuns would have regarded as sacrilegious: evidence that even human spirituality and religious piety might sometimes be nothing more than mating strategies.

Chapter 10

SEX AND RELIGION

It's 1958. Dion and the Belmonts are doo-wopping in the background, and a working-class Italian American man walks by the corner of Forty-third Street and Thirtieth Avenue in Astoria, Queens. As he passes in front of an old Catholic church, the man makes an unmistakable sequence of gestures, waving his hand rapidly from his forehead to his chest and then from his left shoulder to his right. I am describing a scene from a movie starring Robert De Niro, but when I saw that scene, it triggered a cascade of profound memories and emotions in me. During the real 1950s, I made that same sequence of gestures as I passed that exact spot every day, mumbling automatically under my breath "name udda Fathuh, n'uv da Son, n'uv da Holy Ghost, Amen." Every Sunday, inside that very church, bathed in light filtering through stained-glass windows, I would kneel along with elderly grandmothers who had immigrated from Italy and Ireland, stare up at a statue of the Blessed Virgin standing on the head of a snake, and listen to a priest dressed in brilliant vestments chanting, "In nomine Patris, et Filii, et Spirutus Sancti . . . ," often with a thick Irish brogue. Each weekday, I would spend seven hours in St. Joseph's school, a building attached to the church, where Dominican nuns would test my knowledge of passages from the Baltimore Catechism.

Q. What is man?

A. Man is a creature composed of body and soul, and made to the image and likeness of God.

Q. Who is God?

A. God is the Creator of heaven and earth, and of all things.

Q. How shall we know the things which we are to believe?

A. We shall know the things which we are to believe from the Catholic Church, through which God speaks to us.

The Catechism contained a prayer called the Apostle's Creed, which we were told embodied the chief truths of the Roman Catholic Church:

I believe in God, the Father Almighty, Creator of heaven and earth; and in Jesus Christ, His only Son, our Lord; who was conceived by the Holy Ghost, born of the Virgin Mary, suffered under Pontius Pilate, was crucified; died, and was buried. He descended into hell: the third day He arose again from the dead: He ascended into heaven, and sitteth at the right hand of God, the Father Almighty; from thence He shall come to judge the living and the dead. I believe in the Holy Ghost, the Holy Catholic Church, the communion of Saints, the forgiveness of sins, the resurrection of the body, and the life everlasting. Amen.

As a devout young Catholic boy, I learned that if I doubted any of that, I would be branded a heretic and thus cast into the eternal fires of hell. But sometimes it was hard not to wonder about the logic of an all-loving God who would create a world peopled with starving children, who might later be subjected to eternal torture if they succumbed to logical doubts or if they happened to be born in any one of the world's non-European countries, where almost everyone believed in a different version of God. The nuns would get annoyed when my friends and I raised these issues in class, and at least one of the good sisters complained to my mother about it. Despite the nuns' concern, though, I remained a believer during my years in St. Joseph's school. Later, though, after I had been expelled from two Catholic high schools, I was cast into a public educational system populated with Protestants, agnostics, and nonbelieving Jews. Whenever religious beliefs came up in conversation, my new friends chipped away at what I had previously regarded as a few minor logical cracks in the foundations of the Roman Catholic worldview.

As a young man, I had two alternative stories about why I stopped going to church. One focused on the time I attended Mass during my early college years and heard a priest railing against draft-card burners. Regardless of one's political opinion about Vietnam, I said to myself, it seemed the ultimate hypocrisy for a man who supposedly represented Jesus Christ to use the pulpit to support a war. So I left the church for reasons having to do with high-and-mighty moral principles. Or at least that was what I used to tell myself.

Of course, faithful Catholics watching this movie-plot version of my life might doubt my story. If I really were being logical, I could have simply dismissed this sermon as the failing of one particular priest. Hadn't the nuns at St. Joseph's taught me again and again that one of the Ten Commandments was "Thou shalt not kill"? Why should one priest's hypocrisy turn me against a set of doctrines that supported universal love and brotherhood?

I had an alternative version of my falling away from Catholicism, though, one that I would tell my friends after a couple of beers. That one went like this: If there was anything the nuns had really drummed into our heads, it was that sex was sinful. There could be exceptions to the rule against murder—if you were fighting against Hitler, for example, or defending nuns from Mau Mau terrorists—but there was absolutely no exception to the rules against premarital sex, masturbation, and all the other temptations of the flesh. Yielding to any of these temptations was classified as a mortal sin. Likewise sinful were any willful mental meanderings—such as looking longingly at the half-naked women in the erotic magazines hanging temptingly on the rack in the neighborhood candy store. (For non-Catholics who might not know its significance, a single unconfessed mortal sin is considered a one-way ticket to hell.)

At the time I heard the war-mongering priest's sermon, I was eighteen years old and beginning my first long-term romantic relationship. I was therefore spending the majority of my waking hours having thoughts, and when lucky, engaging in actions, that qualified me for eternal damnation. So version two of my falling away from the church was this: I no longer wanted to feel guilty about sex, so I decided it was time to choose another set of beliefs.

Which is the true story? Was it philosophical and logical inconsistencies that inspired that young Catholic boy to heresy, or was it the temptations of the flesh that led my mind to recompute the relative payoffs for blind faith? Of course, that question is impossible to answer for any particular case, even one's own (or maybe particularly one's own). But my colleagues and I have since done research that reflects on the general question.

The Psychology of Belief and Disbelief

During the last few years, there has been a surge of research interest in the psychology of religion. The increased scientific interest in heavenly cognition was probably stimulated by two factors. One is the fact that religious beliefs are intimately tied to the seemingly endless conflicts in the Middle East, where massive carnage has often been couched in terms of Muslims versus Christians, Muslims versus Jews, Muslims versus Hindus, Sunnis versus Shias, and so on. The second is the fact that in the United States, the so-called Religious Right started a new wave of lobbying against the teaching of Darwin's theory of natural selection in the public schools. Former president George W. Bush and vice-presidential candidate Sarah Palin, neither of whom wins high respect in scientific circles, went on record as considering "intelligent design" (aka biblical creationism) a legitimate "scientific alternative" to Darwin's theory. Supporters of creationism want more than the freedom to advocate their religious beliefs in church; they want to make it a requirement that their religious beliefs become a part of the public school curriculum. The political threat of allowing religion to dictate the teaching of science led a number of scientists to subject those religious beliefs to the same sort of critical scrutiny to which scientific claims are regularly subjected.

Evolutionarily oriented psychologists, biologists, and anthropologists interested in religion have focused mostly on questions about the genesis of religious beliefs. Psychological anthropologist Pascal Boyer has suggested that several different mental modules might underlie different aspects of religious belief. For example, religious rituals might be governed by the same regions of the brain involved in compulsive avoidance of filth and disease, which, when overactive, can result in obsessive-compulsive neuroses. Anthropologist Scott Atran has joined social psychologist Ara Norenzayan to consider how the mental inclination to seek causes for natural events might underlie religious beliefs in invisible causes (for example, when Mom comes home and finds a broken dish, she assumes someone is responsible even though she did not witness the act; likewise, when a natural disaster knocks out a village, maybe some unseen being was angry at the villagers). Psychologist Lee Kirkpatrick has suggested that religious beliefs about powerful fathers and loving families might be by-products of brain mechanisms involved in forming attachments between parents and children.

Another approach asks not about the origins of religious beliefs but about the possible adaptive functions of belonging to a religious group. For example, biologist David Sloan Wilson analyzed religious groups as a tool for group selection, arguing that groups that can mobilize their members to share with one another and that can band together to protect themselves against other groups are more likely to persist than aggregations of self-centered individualists. And other researchers, including Azim Sharif, Ara Norenzayan, and Dominic Johnson have combined the two approaches, asking how belief in an omniscient God might inspire people to follow a group's rules and act generously when no one is watching.

Jason Weeden, Adam Cohen, and I approached religion from another adaptationist angle. Rather than searching for the causes of heavenly beliefs inside people's heads, we started the search inside their bodies, investigating how religious participation might directly serve some people's reproductive strategies.

Reproductive Religiosity

Adam Cohen and Jason Weeden both came to ASU in 2006. They both had Ph.D.'s in social psychology from the University of Pennsylvania, and they both had interests in evolutionary psychology. Despite those similarities, they could not have been much more different. Cohen is from Philadelphia, and his central self-defining characteristic is his Jewish identity. Not only does Adam have a program of research on the psychological differences that distinguish Jews from Protestants and Catholics, but he carries on a continuous Henny-Youngman-meets-Woody-Allen comedy routine and peppers his sentences with Yiddish words like "schtupping," "schmuck," and "schicksa." Weeden, on the other hand, speaks with a slight middle-Texas drawl and claims to have rarely been inside any house of worship. Jason also has a law degree from the University of Texas, and when he finished his Ph.D. in psychology, he was offered a job as a corporate lawyer. After three years of wearing a suit and tie, Jason made enough money to take some time off and pursue his interest in evolutionary social psychology.

Weeden was fascinated with the psychology of religion and politics, and he had a very intriguing idea he wanted to investigate. He believed that much of the fighting between America's Religious Right and Liberal Left was based not in a disagreement about high-minded philosophical ideals but in something much simpler and far less noble: The two camps are playing out fundamentally different mating strategies. And they do not like each other because the people playing one mating strategy are actively getting in the way of those playing the other mating strategy. Because I had done research on mating strategies and Adam had done research on the psychology of religion, Jason recruited us to work with him to test this idea.

Weeden pointed out that the United States is often viewed as a highly religious nation, at least in comparison to other Western countries. But Jason loved to play with giant survey databases, and he pointed out that the U.S. population is in fact remarkably divided in its religiosity. According to data he gathered from the 2006 wave of the U.S. General Social Survey, 40 percent attend services several times a month, but 42 percent of American adults hardly ever attend religious services. What are the causes and consequences of the rift between the hyperreligious and the irreligious? Jason proposed what we later came to call the *reproductive religiosity model*. On this view, a primary function of religious groups in the contemporary United States is to bolster a monogamous reproductive strategy, characterized by low promiscuity, exclusive heterosexuality, and a high value on marriage and fertility. Religious groups bolster this reproductive strategy in two ways: On the negative side, they enforce a set of strict moral norms (treating sexual promiscuity as sinful); on the positive side, they provide various forms of support for families that live according to those rules.

Treating premarital sex as sinful provides an incentive to marry early. And treating abortion and birth control as sinful encourages people to have children. According to Weeden, this helps explain why the rank-and-file members of the Religious Right tend to be less well educated than their counterparts on the Liberal Left. Taking care of a family makes it difficult to stay in school to pursue an advanced degree in philosophy or neuroscience. This is one of the trade-offs involved in adopting the monogamous, long-term, high-fertility strategy. From a purely reproductive perspective, this strategy also has costs as well as benefits. For monogamous family-oriented men, the high level of investment in their wives and children means forgoing other mating opportunities. Because a man can never be completely sure that he is the father of his wife's children, the strict religious rules against promiscuity help him too, by reducing the risks of paternal uncertainty. Monogamous, family-oriented women strike a complementary bargain: The strictly enforced norms help keep their husbands from running around, but they also reduce the woman's opportunities to run around with a charming unrestricted guy who could provide sexier genes for her children. (Researchers including Steve Gangestad, Randy Thornhill, and Martie Haselton have found that when women in relationships are ovulating, they become more attracted to guys who look like Vince Vaughn and George Clooney, especially when the women are married to average-looking guys.)

By checking up on one another, the members of a sexually conservative religious community reduce the potential costs of early marriage and high familial investment. Religious communities don't just frown on promiscuity; they condemn it, and they impose costs on those who break the rules, impugning their reputations and ostracizing them. But besides punishing sexual promiscuity, religious groups reward family orientation in various ways—they set up preschools, they join together to share babysitting chores, and they provide assistance to their members when they lose their jobs or get sick.

The prototypical member of the Liberal Left, on the other hand, plays a very different life strategy. He or she may wait until at least the end of college before marrying and beginning to have children and then may delay even a few years longer to go to graduate school, law school, or medical school. Because the human ability to resist sexual urges has a hard time outlasting all that postponement, these folks do not like the Religious Right's attempts to impose rules against pre-marital sex, nor do they like anything that limits their access to all the tools of family planning. Liberal Lefties typically do not give a hoot what you do in the privacy of your bedroom or whom you do it with. These folks pose a problem for the Religious Righties, though, be-cause a large number of sexually loose young people playing the field threatens to disrupt the strict system that religious folks have set up to enforce and reinforce family bonds.

To check the validity of his reproductive religiosity model, Weeden analyzed a mountain of data—from 21,131 people in the U.S. Gen-eral Social Survey. Cohen and I also joined him to look at data from another more focused survey, in this case of 902 undergraduate stu-dents at four American universities who were asked questions about their family plans, their sexual attitudes, their religious attendance, and their moral feelings about stealing, lying, and so on.

We found that people's inclination to attend religious services could be predicted by some of the usual variables—being a woman, being older, being a nondrinker, and being low in sensation-seeking and high in conscientiousness, for example. We also found that attending church was linked to people's opinions about nonsexual transgressions like lying to parents, shoplifting, cursing, and using illegal drugs. But what was more interesting was this: The strongest predictors of attending church were those related to sexual and family values (opposition to in-fidelity, to premarital sex, to abortion, and so on). And when Weeden statistically controlled for the sexual and family value items, the links between religious attendance and the other variables disappeared.

These findings make two important points. First, conservative attitudes about sex and reproduction are at the heart of people's participation in traditional religious groups. Second, attitudes about sex and family are causes of religious attendance, and not just the effects of religious training. The traditional view was that religious teachings cause people to hold conservative attitudes about sex; our findings suggested a very different causal path: that conservative attitudes may cause people to become involved in religion. If this view is correct, people may decide to increase or decrease their level of religious participation as a function of whether that participation advances or hinders their current sexual and reproductive strategies.

Back to my own falling away from the Catholic Church. It happened at a time when I had started attending college and was beginning to realize that I would be in school for a lot longer if I wanted to become a research scientist. Although I was not planning on getting married soon, I was not doing a very good job of avoiding the temptations of premarital sex. So my early theory that I was forced to choose between sex and God might apply to many college-educated people, and not just Roman Catholics. Weeden also has some evidence that many university students switch away from traditional religious beliefs during college, when the temptations of uncommitted premarital sex combine with obstacles to getting married. Later on, when the college-educated settle down to family life, many of them switch back to traditional religious beliefs. Again, people's love lives may drive their religiosity at least as much as their religiosity drives their love lives.

How Flexible Is the Link Between Religiosity and Reproduction?

We began to wonder whether the link between religiosity and reproduction was malleable enough to be moved around with a laboratory manipulation. Could we make people more or less religious just by having them think about attractive mates, for example? Weeden had his doubts, and they were well-founded; although it made sense that someone could undergo a gradual change in his or her life strategy if there was a big change in his or her mating opportunities, that shift should not happen in a few minutes.

On the other hand, people can only change if they have mental mechanisms that calibrate themselves to the current environment. As I described in Chapter 2, for example, we have done laboratory experiments showing that people's commitment to their long-term partners can be shifted merely by informing them that there are a lot of available and desirable members of the opposite sex around. And we have also found that people's opinions of their own mate value could be moved around in a short experiment by telling them that there are a lot of available and desirable members of their own sex on campus. These findings told us that events in the short-term can kick in some of the same shifts in mating strategy as do long-term changes in the environment. If religiosity is to some extent a mating strategy, then it might well respond similarly to information about the local mating pool.

To investigate this possibility, we began a series of experiments with Jessica Li, who was a bright and eager new graduate student working with Adam Cohen and me. In these experiments, we brought students into the lab and told them the study was about evaluating dating profiles. If you were a participant, you would have been told:

> Many students at ASU come from far-away places, like Los Angeles, New York, or Chicago. They are often interested in meeting other people, but they feel uncomfortable going to bars and meeting total strangers. So the ASU student government is trying to set up a computer-based system to allow students to meet one another in a more comfortable way.
>
> The psychology department is helping to set up the system because psychologists know about survey design and social behavior.

The key thing we are interested in here is whether this is the right information and whether it is presented well. The information is from people who have signed up to meet other people at ASU.

After reading that, you would have looked over six dating profiles that were presumably from other ASU students. The people in the photos were actually highly attractive models. The photos were accompanied by self-descriptions in which the attractive people talked about their positive characteristics and expressed their eagerness to date. Half of the time, subjects saw attractive and available people of their own sex; the other half of the time, they looked over six attractive and available people of the opposite sex.

When you were done rating all the profiles, you would have been told we needed some information about you. At this point, you would have answered some questions about your own attitudes on several topics. Buried in the list were questions about the extent to which you believed in God and about whether you thought people would be better off if religion played a bigger role in their lives.

Before we ran our study, we were not really sure what to expect. I thought that perhaps seeing beautiful available women might have made men less inclined to be religious. As it turned out, however, seeing attractive people of the opposite sex had no effect on either men or women. Instead, we found that looking at attractive people of one's own sex led both men and women to express more belief in God.

Why? We think that the results fit with Weeden's ideas about dueling religious strategies. When you become aware that there are a lot of highly attractive mating competitors out there, it reduces the perceived benefits of playing a fast and loose mating strategy (a strategy that is popular among many undergraduates at schools like ASU, where mating opportunities sometimes seem unlimited). For women, a lot of attractive competitors means less attention from the attractive men who might provide good genes, and fewer fellows vying to support your offspring. For men, on the other hand, an abundance of especially handsome and available guys means that if you are playing the field, you may be playing with yourself for most of the game. Under circumstances of limited opportunities, any normal person—who does not look like a fashion model—could benefit from the religion-based supports for monogamy.

Psychologists have traditionally focused on the ways in which early religious indoctrination might lead people to later shun sexuality. William H. Masters and Virginia E. Johnson, the famous sex therapists, listed religious training as one of the big causes of sexual inhibitions. It is certainly true that many religions teach young people that premarital and extramarital sex are evils to be avoided. But the results of our studies suggest that the causal arrow may go in the opposite direction as well. Not only can religion shape people's sexuality, but people's sexual strategies can also shape their religiosity.

Zen and the Art of Atheism

The findings I just discussed also support the less flattering version of my life story—the one in which I left the Catholic Church because my choice came down to either devotion to the Virgin Mary or surrender to the recreational sexual opportunities of the 1960s. If so, my principled objection to war-mongering priests was just a self-serving excuse. But besides inspiring a bit of humility, these various findings also helped tone down my self-righteousness about my political beliefs. As a card-carrying member of what Spiro T. Agnew dubbed "the knee-jerk liberal" set, I often find myself ranting and raving about the moral inconsistencies of the Religious Right, the Sarah Palin–supporting, church-on-Sunday-guns-on-Monday crowd. But it turns out that Jason Weeden had it right: A whole lot of the differences between the Left and the Right boil down to dueling mating strategies. Oddly enough, that makes it easier to take a Zen perspective when I see a big SUV sporting a Jesus fish logo right alongside a sticker that says "Support Our Troops."

In this chapter and in Chapter 9, I have implied that some of those behaviors that seem uniquely human—like the desires to build pyramids or care for the poor—are linked to baser motivations and simple selfish biases that we share with other animals. I've reviewed research suggesting that even those creative and spiritual motivations that defined the top of Maslow's pyramid are not divorced from biology at all. All of this implies that the heights and depths of human motivation are intimately connected to one another and that we need to revise our self-perceptions as enlightened beings standing outside of nature. But the news is not all bad. I will now explore the flip side of that equation and present some evidence that many of our judgmental biases that seem hopelessly irrational to economists are, on closer examination, deeply rational.

Chapter 11

DEEP RATIONALITY AND EVOLUTIONARY ECONOMICS

Almost four decades have passed since I was that long-haired graduate student who walked into a bookstore to avoid studying for his doctoral qualifying exams. My research adviser from those days, Bob Cialdini, was also a young guy, fresh out of graduate school himself. But a few months ago, I introduced Cialdini's "last lecture" to our department on the occasion of his retirement.

Although I still like to think of myself as a young rebel, I now see a "distinguished" gray-haired fellow looking back at me when I pass by the mirror. (I don't look there so much anymore, unless I'm trying to trim the distinguished white hairs that now sprout randomly from my ears.) And some of my other once-youthful graduate student fellows, the ones who sat with me on campus ogling beautiful young hippie women in halter tops, have also retired. But not me. My retirement account is so thin that if I were to quit working in the next few years, I would either have to pack up and move to Ecuador or turn to a life of panhandling ("will lecture for food").

IRAs Drained by DNA

Although I have underfunded my retirement account, I've spent well over half a million dollars on my two sons. From the perspective of classical economic theory, this feature of my money handling has been decidedly irrational. In the case of my older son, Dave, to take just one example, I could have insisted that he attend ASU, where the dean of the honors' college offered him free room and board on top of free tuition. Instead, I agreed to shell out tens of thousands of dollars so that he could attend New York University's film school (itself a decidedly irrational decision from an economic perspective). Since then, I have given him additional tens of thousands of dollars to attend graduate school, then to make a down payment on a house, and now to help out with the expenses of having two children of his own. (Like many of his freelancing fellow film-school graduates, he has yet to achieve an income that will allow him to buy the mansion next to Steven Spielberg's.)

Besides all this, I have contributed many of my limited nonfinancial resources, spending many hours helping Dave care for his own children. (With another young child of my own at home, I realize how energetically expensive it is to respond to their incessant demands.) In fact, I will have to stop writing this very afternoon because Dave is about to bring the grandchildren over. Time is of course money; during those many hours spent with my own young child and the grandchildren, I could have been tending to retirement investments or doing other things to make more of the long green.

How do I feel about all that spent money and time? From the purely rational economic perspective, I should be sending both sons a monthly bill and making angry phone calls when they fall behind on their repayment plans. The little one is five years old, and has his own bicycle, so perhaps I should soon consider insisting that he take a paper route!

But although I would surely feel great resentment toward a colleague or friend who had run up a half million dollars on my account, I do not feel anger or resentment toward my sons. Instead, I feel guilty that I cannot do more. In fact, I feel bad about even writing this, because I do not want my sons to read it later and worry that I think of them in terms of "economic investments." My two sons instead elicit my warmest and most loving feelings. Irrational economic investments though they might be, I cannot get enough of them.

Besides my "irrational" contributions to my children, I have also given thousands of dollars to the Sierra Club, the Nature Conservancy, the Brady Center to Prevent Gun Violence, and on and on. Just this week, I sent a check to an organization fighting to convince the U.S. Congress to pass a bill ensuring universal health care for other Americans. (I already have adequate health care, and like many of those old bastards who have so successfully opposed "socialized medicine" for young people, I will be eligible for Medicare in a few years.) And I have been trying to teach my younger son to be more generous toward others. Just yesterday, I gave him a dollar to stuff into the Salvation Army kettle as we walked into a store, where we were going to buy an anonymous Christmas gift for a poor family in his school.

Does all this generosity make me a candidate for canonization? Does it mean that my fistfighting sex-obsessed younger self has magically transformed into a saint? Not quite. Indeed, my apparently selfless behaviors are, from an evolutionary perspective, at least as self-serving as my seemingly selfish ones. And yours are too. How this all works begins to make sense in light of some exciting recent developments linking evolutionary psychology to classical economics.

Economic Selfishness, Psychological Irrationality, and Deep Rationality

On the classic model of rational man, we humans are reasonably well-informed decision-makers who make choices designed to maximize our "utility," or expected satisfaction. The general model does a great job of explaining things like supply and demand and of helping us understand how the conflicting selfish tendencies of buyers and sellers result in reasonably priced consumer goods in the marketplace. And it has been useful for economists to think in terms of a common coin of utility; it allows them to compare the psychological value of desirable outcomes as different as a tasty meal with friends, a romantic vacation, and a Porsche Carrera GT.

Over the last few years, a new set of ideas has changed the face of economic theory, with the development of the field of behavioral economics. Challenging the classic model of rational man, behavioral economists have incorporated the insights of cognitive and social psychology—fields in which researchers have done a rich business demonstrating people's tendencies to use simplistic and irrational biases. In their popular book *Nudge*, for example, economists Richard Thaler and Cass Sunstein suggest, "If you look at economics textbooks, you will learn that *Homo economicus* can think like Albert Einstein, store as much memory as IBM's Big Blue, and exercise the willpower of Mahatma Gandhi." Thaler and Sunstein distinguish between "Econs," those classically rational individuals who make deeply reasoned decisions after a consideration of all relevant sources of information, and "Humans," who make rather less omniscient decisions informed by limited cognitive heuristics and various irrational biases.

Perhaps the quintessential example of behavioral economics is Daniel Kahneman and Amos Tversky's demonstration of "loss aversion"—the finding that people are more psychologically moved by a loss of $100 than by a gain of an identical amount. To a rational economic mind, $100 is worth exactly $100. But Kahneman won a Nobel Prize for a body of work illustrating that this seemingly simple and rational equation ain't necessarily so. Along with Tversky, Kahneman demonstrated in various experiments that losing $100 has more psychological impact than gaining $100 and that the typical human will pay more to ensure that he or she will not lose $100 than he or she will pay for an equal chance of getting another $100.

Econs, Humans, and Morons

The infusion of cognitive psychology had a jolting impact on how we think about economics. I predict that adding evolutionary psychology into the mix will produce yet another seismic uplift in the terrain. I would argue that behavioral economists are only partly right in their focus on the limitations of human decision-making. Although I buy the distinction between Econs and Humans, I would distinguish between the behavioral economic view of "Humans as morons" and the evolutionary psychological view of "Humans as clever apes" (or let us call them "Evols"). Behavioral economists have focused on the proximate biases involved in heuristic decision-making, and I do not dispute that people often make quick-and-dirty decisions. However, the emerging evolutionary view is more in line with a position advanced by Gerd Gigerenzer and Peter Todd and their colleagues: that human beings use simple heuristics in ways that make us surprisingly smart.

In previous chapters, I mentioned three of my former graduate students—Jill Sundie, Norm Li, and Vlad Griskevicius—each of whom had studied economics before shifting their interests to evolutionary psychology. After thinking about the connections between ideas from economics and those from evolutionary psychology for several years, we have recently joined forces with Steve Neuberg and Jessica Li to advance the argument that many of people's seemingly irrational choices are actually manifestations of what we call deep rationality.

On the deep rationality view, decision-making indeed reflects psychological biases, but those biases are anything but arbitrary and irrational. Instead, they are the outputs of mental and emotional mechanisms designed to maximize not immediate personal satisfaction but long-term genetic success. Furthermore, our view incorporates the notion of modularity in a central way. Instead of a single rational decision-maker operating according to a single set of utility-maximizing rules, this view presumes that all of us have a number of different economic subselves inside our heads. Each of our economic subselves pays attention to different costs and benefits and weighs them differently, in ways designed to deal with the most prominent threat or opportunity on our life's horizon.

How to Make the Prisoner's Dilemma Disappear

If you have ever taken a course in either economics or social psychology, you have heard about the "prisoner's dilemma." Here is how it goes: Imagine you are one of two crooks who have been arrested while trespassing at the scene of a potential heist, and that you are being held on suspicion for a string of burglaries. You have two options: remain silent (thereby *cooperating* with the other crook in evading prosecution) or confess to the district attorney (thereby *defecting* on your pact of silence with the other burglar). If only one person confesses, thereby providing the district attorney with solid evidence against the other, the one who confesses goes free. For the pair of you, the best outcome is if you both cooperate in remaining silent (in which case you will both get a short sentence for trespassing). But the decision poses a dilemma: If you remain silent and the other crook confesses, things will turn out really badly for you. From a purely economic perspective, the game is rigged so that the most rational decision is to defect. If the other crook keeps quiet, then you go free; if he defects also, you get a lighter sentence than if you had kept quiet. But it is a dilemma because, if both individuals make the purely selfish decision, they fare more poorly than if they had acted cooperatively.

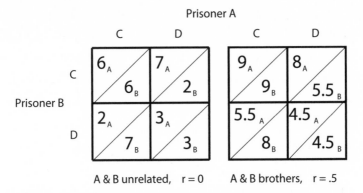

FIGURE 11.1 A disappearing dilemma. The standard prisoner's dilemma is shown in the box on the left. In each box, the payoffs for prisoner A are above the diagonal, and those for prisoner B are below. For prisoner A, the best payoff always comes from defecting (choosing option D) rather than cooperating (choosing option C) (7 versus 6 if B cooperates; 3 versus 2 if B defects). The figure on the right shows that things change if the two players are brothers, who share half their genes. In that case, each prisoner's payoffs can be recalculated to include half his brother's gains. For the recalculated payoffs on the right, the best choice for each player is to cooperate, regardless of his brother's choice.

If you think in evolutionary terms, however, such dilemmas often disappear when the other player in the game is your brother or your child. Because you share half your genes with your child, your bank of inclusive fitness points gets to add half of your child's gains to your gains (and also to subtract half of his or her losses from your losses). In other words, the evolutionary banker in my head gives me a 50 percent rebate for every dollar I spend on a brother or a son. Figure 11.1 shows how that works for one set of typical payoffs. It illustrates how a situation that would otherwise be a dilemma disappears if the players are close genetic relatives. Federico Sanabria, Jill Sundie, Peter Killeen, and I have explored how this works in more detail, and the bottom line is this: There are still some situations in which it pays for someone to defect on a close relative, but they are a much more limited set than those in which it pays to defect on a nonrelative. All the

standard economic models, of course, implicitly presume you are dealing with a nonrelative and can therefore ignore his or her payoffs and
focus exclusively on your own.

Like a brother, my child shares roughly half of my genes. But there
is another level of evolutionary math that makes it even easier for me
to spend freely on my child. If my genes calculate the value of investing another dollar in my own mating success, it is not as likely to yield
the same marginal utility as those genes would get if I invested that
same dollar in one of my children. I already had sufficient resources
when my son Dave went to college, so rather than investing more in
myself, it made more sense to invest in him going to New York University, where he could develop his artistic film production skills instead of having to work in a factory. As a consequence, he was a more
desirable fellow, who was able to attract college-educated women, and
to produce two grandchildren with high IQs. My younger son, who is
likewise turning out to be a highly intellectual guy, will also probably
benefit from my continuing to invest in him as he grows up and develops his creative talents. My genes would gain no benefit if I made
the more "selfish" decision to stop investing time and money in my
sons and grandchildren, so that I could instead retire early, buy a fancy
recreational vehicle, and learn to swing a golf club.

Life as a Bank Account

All this makes sense in terms of what evolutionary theorists call life
history theory. According to this theory, which I discussed briefly in
Chapter 7, any animal's life can be divided into two or three main
phases, each one defined by a different set of investment trade-offs.
The earliest phase is called somatic effort—when the animal invests
mainly in building and maintaining its own body. The second phase
is mating effort—when the animal invests time and energy into finding a mate. For some species there is (as we noted) a third phase, par-

enting effort—when the animal shifts its time and energy investments toward raising its young.

At what age should an animal shift from somatic investment to mating effort? Should the animal waste time and energy on parenting effort? These are questions about economic trade-offs, and the answers depend on the expected payoffs. Young mammals are notoriously expensive to raise. They need a mother who is willing to carry them inside her body and then allow them to continue sucking out calories for months or years after they are born. So a typical mammalian female would gain nothing if she began reproducing too early, because until she has developed the physical capacities and behavioral skills to store enough calories for two, her young would not survive. Although a male could produce sperm much earlier, males typically wait even longer to switch to mating effort—until they have built a large enough body to effectively compete with the other, more mature males (on the efficient rule, Do not start competing until you have a reasonable chance of winning).

It all works as expected in our species. Human females typically are not capable of bearing young until they have reached their early teens. Boys typically do not reach puberty until a year or two later. As I discussed in earlier chapters, our species is one of a small minority of mammals in which the males do pitch in to help raise the offspring, because human babies develop slowly and have large, energy-hungry brains that make them especially needy.

From a life history perspective, early development—the somatic phase—is, as I noted in Chapter 7, akin to building a bank account. Reproductive effort is akin to spending that bank account. Different animals' life histories involve evolved strategies for spending that bank account on a schedule that will maximize genetic success, given the problems and opportunities that particular animal is likely to encounter. Thinking in terms of life history investments, splurging one's retirement money on a good time instead of reinvesting it in one's grandchildren is only "selfish" at a proximate level. At an ultimate (evolutionary) level, our genes would regard it as self-defeating. So this perspective shifts the understanding of "rationality" from immediate personal payoffs to the much longer-term gene's-eye view.

Shifting Priorities

What counts as evolutionary rationality varies with the species. Elephants take decades to reach sexual maturity; some small mammals can successfully reproduce after only a few months. What counts as an evolutionarily rational strategy also varies within a species and depends on the animal's sex as well as on his or her phase of the life span. And at any given time for any given animal, the immediate environment may offer certain opportunities (or threats) and not others. As I discussed in previous chapters, our minds are equipped with a set of subprograms designed to focus our attention and our mental abilities on the current most-important set of opportunities and threats. The result is that rather than having brains with a single "rational" information processor crunching information, we have a set of different subprograms in there, each crunching information in a way designed to solve the most immediately important problems. Remember one of the key messages from the fusion of evolutionary psychology and cognitive science: The human mind is not a massive information-crunching computer but a multitude of miniminds, a collection of independent mental adaptations specifically designed to solve particular adaptive problems by crunching different kinds of information in very different ways. This has important implications for economic decision-making; what counts as a good decision about allocating resources to solve one problem may count as a very bad decision for solving another problem.

Extending the logic of Chapter 6, we can say that each individual decision-maker has several different economic subselves, and which subself is in charge right now depends on what adaptive threats and opportunities are currently prominent in the environment. What looks like irrationality to one subself may be deeply rational to another. Your marketplace subself, which is dominated by the question, "What's in it for *moi?*," would be aghast at the exorbitant bill your parental subself has run up sending junior through college, for example.

The implications of the evolutionary multiple-mind viewpoint for everyday economic decision-making are thus far largely unexplored. But in this book, we have already covered evidence suggesting that the decision rules a woman uses when thinking about how to negotiate with a stranger in the marketplace will not follow the same math as the mental rules she would use in deciding how to exchange resources with her son, who shares half her genes, and is dependent on her generosity if he is himself to survive to reproductive age. Besides the different mental rules for dealing with strangers and close kin, everyday people need yet another set of decision biases for interacting with friends, to whom they are linked not by shared genes but by trust-based reciprocal exchanges. And romantic partners do business according to still another set of decision rules.

In studies I discussed in earlier chapters, my colleagues and I found that whether or not a person chooses to conspicuously and wastefully throw around his or her wealth, to display his or her benevolence and nurturance, to risk a fight, or to go against group opinion will ebb and flow in predictable ways depending on whether that person is a she or a he and on whether he or she is in a mating frame of mind, as opposed to thinking about status or worrying about life and limb.

The logic of deep rationality suggests that fundamental biological motives such as mating and self-protection should drastically change all the traditional behavioral economic biases, such as temporal discounting (the tendency to value small immediate payoffs over larger delayed ones) and probability discounting (the tendency to prefer a smaller payoff now over a larger one later). The same motives should also change what a person regards as luxury versus necessity, and they should do so very differently for men and women. A series of experiments by Norm Li and his colleagues found profound differences between men and women in distinguishing between luxuries and necessities in different aspects of social decision-making. For example, if people are given an abundant budget of "mate dollars," both men and women choose similar partners: They want someone who is physically attractive, funny, warm, and high status. But most of us are not like a wealthy movie actor who can "have it all"; instead, we have to prioritize. When men are put on a more limited budget of mate dollars, they spend first on physical attractiveness, indicating that that is a high priority. Women on a limited budget make different choices, placing higher priority on getting a partner with sufficient wealth or status and treating good looks as an expendable luxury.

Loss Aversion by Morons Versus Loss Aversion by Evols

Let's reconsider the classic case of "loss aversion" in light of the notion of deep rationality. As I mentioned earlier, the Kahneman and Tversky loss aversion function simply indicates that a loss of a given size (say $100) has more psychological impact than a gain of the same size. This function has now made its way into introductory economics textbooks and has been amply supported by research. Indeed, one recent review concludes, "There has been so much research on loss aversion that we can say with some certainty that people are impacted twice as much by losses as they are by gains." But why the bias, and is it the same for every type of gain and loss?

Evolutionary theorists, including E. O. Wilson, have suggested a possible answer to the "why" question: Ancestral humans would have survived better if they put a higher priority on avoiding losses than on acquiring gains because they frequently lived close to the margin of survival (extra food would be nice, but insufficient food could mean death). Consistent with this idea, loss aversion has been found not only in humans but also in other species (whose ancestors, like ours, would have suffered more from falling below the line of subsistence than they would have profited from an overabundance of resources). This is a plausible functional hypothesis about past conditions, but it does not fully exploit the scientific strengths of the modern evolutionary approach, which we can use to generate specific new hypotheses about when and how loss aversion should ebb and flow with functionally important motivations.

For example, the usual inclination toward loss aversion should be erased or even reversed when a mating motive is activated. Furthermore, this erasure should occur only for males, and not for females. Why? As I noted earlier, women, as female mammals, have an intrinsically high minimum investment in their young, and this inspires them to be relatively more selective in choosing mates. As a consequence, males must compete to be chosen as mates. I talked about various ways for a male to say, "Pick me! Pick me!" One is to flash a noticeably wasteful display (such as a peacock's feathers or a Porsche Carrera GT); another is to directly outcompete the other males (butting them with antlers or winning a fight for a well-appointed executive office). To beat out the competition, it helps to take risks, and as noted in Chapter 9, male mammals indeed become especially risky during the mating season. It follows that men primed to think about mating should act like bighorn sheep during the rutting season, when too great an aversion to losses would prevent the kind of risky competition that can beat out the other males.

If our logic is correct, though, the shape of that famous function should change in predictable ways for men under the influence of a mating motive. Men primed to think about mating ought to shift their attention toward gains and away from losses. Along with Jessica Li, Vlad Griskevicius, and Steve Neuberg—and funded by a generous grant from the National Science Foundation—I set out to test this hypothesis. In one experiment, we had some people (in the mating condition) think about a romantic first encounter with someone they found very attractive, whereas others (in the control condition) thought simply about organizing their desk.

After the motive manipulation, all our subjects answered a series of questions like this: Imagine that you are at the 50th percentile of financial assets (in other words, half the population makes less money than you, and the other half makes more). How happy or unhappy would you be if you dropped to the 40th percentile? (Imagine an 11-point scale, with 1 being extremely unhappy, and 11 being extremely happy.) What about if you went up to the 60th percentile? How happy or unhappy would that make you?

The results came out exactly as we had expected: In the control condition people were inclined to be loss averse—that is, they expected their happiness level to change more after a loss than after a gain. A mating motive had no effect on women's responses, but it did have an impact on men's judgments. Men in a mating frame of mind focused more on the gains, and reduced their normal oversensitivity to losses. That is, their psychological evaluations changed in a way that would encourage them to be more risky. This finding was not just a general effect of becoming more physiologically aroused, either. In a later study, we also put some people in a self-protective frame of mind by having them imagine a scenario in which someone was breaking into their house. When their night watchman self was thus activated, men, like women, became even more loss averse.

Rationality, Irrationality, and Deep Rationality Revisited

By integrating behavioral economics with evolutionary psychology, I believe we are moving toward a totally new way of thinking about economic rationality. Behavioral economist Dan Ariely has eloquently argued that we are "predictably irrational." But this depiction captures only half of the truth. It is certainly true that people do not systematically calculate all the potential costs and benefits of various alternative choices and then reliably choose the one most likely to maximize their future gains. The behavioral economists are correct in pointing out that we instead use simple heuristics, ignoring a great deal of relevant information and thus making biased decisions that do not incorporate all the available options. At a deeper level, however, our biases reflect the influence of profoundly important functionally relevant motivations. Furthermore, our failures to make simplistically "selfish" choices reflect the powerful influence of a deeper rationality. Rather than being designed to maximize immediate personal reward, many of our choices seem designed to maximize our long-term genetic success.

So far, we have focused mainly on the psychology of the individual. Of course, all the individual biases inside your head and mine are designed to produce effects on the social world—to attract mates, to gain status, to protect ourselves from harm, and so on. And as illustrated by the case of the prisoner's dilemma, your individual decisions and mine are intrinsically dependent on and sensitively responsive to decisions made by the people around us. The implications of this dynamic connection between different people's decisions go way beyond two people negotiating a check in a restaurant or trying to avoid a fistfight, though. In fact, your personal decisions link you into a vast network with strangers halfway around the world, including dairy farmers in the Netherlands, financiers in New York, and world leaders in Washington, Moscow, and Beijing. In the next chapter, we'll explore how all that works.

Chapter 12

BAD CROWDS, CHAOTIC ATTRACTORS, AND HUMANS AS ANTS

Every day before school, my little brother and I donned identical blue pants, white shirts, and blue ties. The uniform marked us as students at St. Joseph's, the Catholic elementary school on the corner. There was also a public school on our block, P.S. 70, but our mother didn't even like us playing in the schoolyard there. What troubled her was not just that the un-uniformed public school kids were rowdier than the nun-fearing youngsters at St. Joseph's, but also that the schoolyard was a hangout for the Garrisons, a gang of teenage hooligans who wore leather jackets and tight jeans held up with thick leather Garrison belts (the source of the gang's name). The Garrisons upset the respectable citizens on the block by fighting, drinking, cursing, and having sex in the schoolyard. As a boy growing up in the 1950s, of course, I thought these James Dean–like characters, with their greased pompadour haircuts and Lucky Strikes dangling from sneering mouths, were kind of cool.

The Forty-sixth Street Boys

By the time I started high school, the Garrisons were gone, several to prison. But there was a new generation of public school hoodlums hanging out in P.S. 70's urine-stinking schoolyard. Besides drinking and fighting, the new troublemakers were reputed to be taking drugs, which included popping "downers," and sniffing glue. Although several members of the new generation of rowdies had been my playmates as children, I joined a different crowd. I started hanging out at a nearby city park with a group of teenagers who had graduated from St. Joseph's and had mostly gone on to Catholic high schools.

Although my parents should have been relieved that I had not fallen in with the glue-sniffers in P.S. 70's schoolyard, they nevertheless disapproved of my new crowd. They had hoped I would hang out instead with my fellow students at Regis, an elite Jesuit school in Manhattan where all the students were supported by scholarships. My new associates, however, who called themselves the Forty-sixth Street Boys, were not Regis intellectuals but mostly students from lower-tier Catholic high schools like Power Memorial and Mater Christi. Some of the Forty-sixth Street Boys had even dropped out of the Catholic school system to attend Bryant High, a city school where students sniffed glue in the restrooms and where a full-time policeman was assigned to roam the grounds. Rather than spending their time in the library and dressing in preppie Brooks Brothers–style outfits, like the budding young scholars at Regis, the Forty-sixth Street Boys dressed in tight sharkskin pants and greaser-style half-boots while they hung around the park smoking cigarettes, flirting with girls, and listening to doo-wop music on transistor radios. I was, as my parents observed, hanging out with the "wrong crowd."

Looking back, I can see that my parents were absolutely right. The Forty-sixth Street Boys were not only a bad influence on me, leading me to sneak out of church on Sunday to smoke Luckies and drink Gallo port wine, but they were seriously rotten friends. They ridiculed me for attending Regis, which they claimed was full of "brown-nosed faggots" (the term then used for intellectual nerds). They made fun of me for being tall and skinny, for having a big nose and big feet, and the bigger ones bullied me mercilessly, then called me a "punk" when I backed away from fistfights with guys like Martie Magno, who was built like Al Capone and who liked to pound his opponents' heads against the pavement to finish up his fights, which he never lost. Although I was at the very bottom of this hierarchy of hooligans, I continued to strive for their acceptance instead of hanging out with other skinny geeks, who actually valued the intellectual pursuits at which I might have excelled. While wasting endless hours at the park enduring ridicule from this pack of bullies, I completely ignored any intellectual pursuits. Instead, I stopped studying and was expelled from Regis (hoping that the tough guys would stop treating me like a nerd if I didn't go to a nerdy school).

My expulsion from Regis was not the end of the downhill slide precipitated by hanging out with the wrong crowd. Within six months, I got expelled from another Catholic high school—Power Memorial—this time for cutting up in class and continuing my moratorium on studying. Fulfilling my parents' greatest fears, I ended up bebopping down the halls of Bryant High in street clothes, right alongside the glue-sniffers. My parents decided to move the family out to Long Island, a migration designed in part to get me away from the neighborhood before I became one of the serious troublemakers and followed in the footsteps of my long-lost biological father, who was at that time residing in Sing Sing.

The Wrong Crowd, Again

Although my family's move to Long Island saved me from one bad crowd, I never fully recovered from my inclination to hang out with troublemakers. I was almost expelled from college after coming to class drunk and heckling my first psychology professor. I had been out drinking in the middle of the day with two of my old greaser friends who were on leave from the navy. I was in a community college, and on academic probation, toying myself with the idea of joining the navy.

Although I did not usually listen to my stepfather, he was successful in convincing me that, given my problems with authority, I would hate the military life. Furthermore, there was a nasty war raging in Vietnam, and several of my old friends had already been killed there. Uncle Sam was still giving college deferments then but was drafting people as soon as they dropped out. That made expulsion from college a possible death penalty. So I got serious about my studies, and I was surprised at how much I enjoyed college courses when I actually did the assigned readings. I began to realize that I was much better suited for the life of the mind than life among the tough guys with tattooed biceps.

But even after I completed my return trip to the world of intellectual nerdhood, I continued to seek the company of rebellious characters. In the academic world, the rebellious crowd tends not to break windows, start fistfights, or beat up old ladies. But they do like to break conventions, start arguments, and overturn old ideas. In the 1990s I came under the influence of two such intellectual badasses. One was Guy Van Orden, a new assistant professor at ASU. The other was Bibb Latané, one of the most successful brainiacs in the field of social psychology. Although Guy and Bibb had not themselves met at that time, they were members of the same intellectual gang, and they both led me astray in the same direction.

Guy Van Orden came to ASU after studying cognitive science at the University of California at San Diego. Like me, he had not come from an academic background. He was from a Mormon family in Idaho, but he didn't fit any of the stereotypes about Mormons. Guy was not the polite, well-behaved, teetotaling young fellow carrying the Book of Mormon to your door. His appearance and behavior were closer to those of a guitar player in an alternative rock band, complete with black sneakers and a ponytail (in fact, one of Guy's subselves was a musician). Guy could drink most Germans under the table, and at departmental parties, he would still be standing, beer in hand, at four in the morning, arguing about philosophical ideas.

Guy's favorite topic to fight about was not religion, drugs, or politics but philosophy. When we talked about the philosophy of science, Guy would attack me with terms like "reductionist" and "determinist," which he used like a born-again Christian would use "sinner" and "heretic." I would shout back at him, "Of course, I'm a reductionist and a determinist! And I can't see why you spit those terms out like insults. It's called science!" To me, opposition to "determinism" was an intellectual cop-out, a ploy used by social constructivists who were too lazy to do actual research and wanted to use big words to mask their unwillingness to look rigorously at the natural world. But Guy was not some lazy intellectual bullshitter; he was a tough-thinking scientist and a fan of an alternative scientific perspective called "dynamical systems theory."

Bibb Latané is a prominent social psychologist who entered the field a couple of decades before Guy Van Orden. Like Van Orden, though, Latané liked to party into the night with a drink in his hand, arguing about ideas. He liked it so much, in fact, that he had converted his massive summer house on the beach in Nags Head, North Carolina, into a conference center, to which he would invite diverse groups of researchers to get together for weeklong sessions talking about ideas and drinking into the night. Latané came from a wealthy family and went on to become one of the world's most influential psychologists, but he nevertheless earned a reputation as something of an intellectual troublemaker. A physics major before switching to psychology, Latané liked to shake up social psychology by bringing in radical new ideas from mathematics and other sciences. In the 1990s, he was, like Van Orden, talking enthusiastically about dynamical systems theory.

Latané had begun using ideas from dynamical systems theory to explain the spread of social influence in crowds, to understand sudden flips in political attitudes, and to elucidate how cultural groups come to share whole sets of apparently random attitudes and behaviors (how, for example, some large groups of white Americans came to wear John Deere caps, listen to Hank Williams, eat grits, get married in Baptist churches, and say "y'all" and "pardon me, ma'am?" while others came to wear Garrison belts and leather jackets, listen to Dion and the Belmonts, eat salami heroes, get married in Catholic churches, and said "youse guys" and "hah?"). As it turns out, Latané's research demonstrated why my parents were right that the crowd you are in really matters.

Chaotic Attractors and the Revenge of the Nerds

I once saw a cartoon depicting a tough-looking teenage boy in a leather jacket surrounded by a group of intellectual nerds dressed in jackets and ties, the sort of brainy kids I met during my brief tenure at Regis. The tough kid is trembling with fear as the nerds taunt him with a volley of questions like "C'mon, Bruno, what's the Pythagorean theorem?" "Define 'arcane,' Mr. Wise Guy!" "What's Newton's second law, Bruno?" When people like Van Orden and Latané started tossing out ideas about dynamical systems theory, I felt a bit like the ignorant leather-jacketed punk in that cartoon.

Whenever Van Orden started ranting about "dynamical systems," he would spew out lots of other scary-sounding terms, like "chaotic attractors," "cusp catastrophes," and "fractals." Although Van Orden left ASU, the university later set up a center devoted to the study of complex dynamical systems, where a group of brilliant biologists, psychologists, economists, and mathematicians now gather to speak that same strange language—using words like "bifurcation" and "hysteresis" in a sentence as easily as if they were talking about oatmeal or bicycles. You have to be careful about asking for a concrete example, because they may illustrate their point by referring to differential calculus or by writing an equation on the board.

Because Van Orden and Latané were so obviously brilliant and so clearly convinced that the new ideas about complex dynamical systems could help us understand almost anything you could think of, I was inspired to read a few popular books on the topic, such as Mitchell Waldrop's *Complexity: The Emerging Science at the Edge of Order and Chaos* and Fritjof Capra's *Web of Life: A New Scientific Understanding of Living Systems*. I began to see how these new ideas could combine with concepts I was already using—from evolutionary psychology and cognitive science—to provide a whole new understanding not only of how the mind works but also of how the simple selfish rules inside our individual heads combine to make families, businesses, governments, and whole societies work.

I am not going to launch into a textbook-style chapter on cusp catastrophes and chaotic attractors. To be honest, I still speak that language like I speak Italian or Spanish, enough to communicate with a five-year-old child who is patient enough to speak very slowly to me. So I only want to talk, at a kindergarten level, about three ideas that are central to this way of thinking, and about how these grand-scale ideas connect with the simple selfish biases I've been talking about in most of this book.

Important Idea Number 1 is *multidirectional causality*, the idea that causes and effects are tough to tease apart in nature because an effect can turn around and alter the thing that caused it. Here's a simple example. If you had a hidden video camera in my house you could occasionally catch me saying to my five-year-old son, "Liam! Stop whining at me and just put on your jacket, or you'll be late for school!" My voice at these times is stern and loud, intended to cause him to stop yelling about a missing Lego part and hustle out the door. But such attempts at influence frequently backfire, causing Liam to yell

even louder and to redouble his efforts to influence me in his preferred direction. If I'm in an especially big rush to get him out of the
house, I may respond by raising my volume and sternness level another notch. That is likely to prompt him to scream even louder and
tell me to shut up, which earns him a time-out, and consequently
makes us even later. All this loud talk may bring my wife into the action, as she attempts to bring peace, but she may find her efforts re-
warded with each of us now loudly resisting her.

As it turns out, all social life is like this; instead of unidirectional
causality, we have complex multidirectional influences. We try to exert
influence on our family members, neighbors, and coworkers, and
those family members, neighbors, and coworkers in turn try to influence us back, and they influence one another, which can indirectly affect us in still other ways.

Because there is so much going on out there, with people influencing one another back and forth, and random new people throwing haphazard new vectors into the continuous stream of multidirectional
influences, you might expect to find that reality would be mostly a
booming and buzzing chaos of disorder. But enter Important Idea
Number 2: Systems theorists have discovered that nature is chock-full
of *self-organization*. Order often emerges spontaneously out of randomness, maintained not by some overseeing governing authority but
by simple, self-centered interactions between local players. Even on the
worst of days, Liam and I manage to get on track and out the door and
to arrive at his school right along with hundreds of other families, each
of whom has managed to settle its own internal conflicts of influence.

Another fascinating conclusion to emerge from research on complex systems gives us Important Idea Number 3: Tremendous complexity can result from just a few interacting variables. As Bert
Hölldoebler and E. O. Wilson point out in their book *Superorganism*,
for instance, ants have tiny brains with only a small array of simple in-
stinctive decision rules, and yet they are able to construct complex so-

cieties with flexibly organized castes dividing up labor, solving diverse environmental challenges, and constructing architecturally brilliant living structures. And as Sean Carroll points out in his book *Endless Forms Most Beautiful*, genetic researchers have been surprised to find many fewer genes than they expected and to discover that the vast majority of them are shared across species as widely separated as cockroaches and human beings. For example, the very same gene that governs the development of an insect's six legs is the one that governs the development of our four limbs. Slight changes arising from the interaction of different genes, however, have profound and complicated consequences.

Self-Organization: Order out of Randomness

While I was hanging out with the greaser crowd, one of the courses I failed to study was algebra, which meant that I did not move on to take calculus. Unfortunately for me, a lot of the folks who are into complexity theory like to express their ideas in equations, to which my brain responds as it does to someone speaking Italian very fast (bringing back images of standing on a train platform in Italy with the speaker rapidly blaring, "*Il treno per Firenze é appena partire al binario due; il treno di Milano é appena arrivare al binario cuatro; il treno di Venezia . . .*").

But fortunately for me, you can understand the concept of self-organization by looking at pictures, with no equations (and no Italian) necessary. In fact, I was amazed when I discovered that I could use the simple spreadsheet on my computer to make self-organization appear right before my eyes. One night, after hearing Guy Van Orden and my colleague Sandy Braver talk about Latané's research on the spread of influence in groups, I went home and drew a simple matrix on my screen, with a salt-and-pepper configuration like the one on the upper left in Figure 12.1.

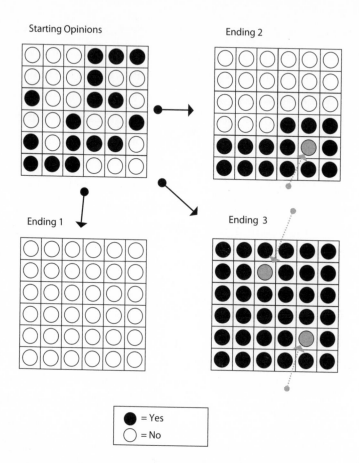

FIGURE 12.1 Self-organization in neighborhoods. The upper-left box depicts a neighborhood in which neighbors are split randomly on a particular issue. Assuming everyone wants to fit in with the majority of his or her neighbors, a few rounds of conversations between immediate neighbors will lead the neighborhood to become uniformly opposed. If one or two neighbors are committed to a favorable position (the grey individuals on the right), the final pattern may be profoundly different (as described in the text).

Think of the picture as a neighborhood, where the neighbors have to decide whether to vote yes or no on a plan for a special tax to build a new school. Let us assume that the initial distribution of opinions is mixed, and neighbors who are pro or con are randomly intermingled with one another at the beginning. Assume also that people will talk to their immediate neighbors, and that they are motivated not to hold an unpopular opinion. After I set up a matrix like the one in Figure 12.1, I then had Excel update each individual's opinion to take into account the average of his or her immediate neighbors' original opinions.

So, after the first round of discussion, the person in the third house down on the left changed his opinion from yes to no, while some of his neighbors changed in the opposite direction. The final pattern was different depending on which random pattern I first plugged in. Nevertheless, after just a few rounds, the salt-and-pepper arrangements disappeared, and the neighborhood always ended up in a simpler, more uniform pattern. In this case, if everyone in the neighborhood simply attempted to bring his or her opinion in line with the majority, this neighborhood would have ended up being unanimously opposed, as in Ending 1 on the lower left.

Systems like this can start out very unstably, and small differences at the beginning can have a big effect on what the eventual equilibrium looks like. For example, Ending 2 (upper right) depicts what would happen if just one person (marked in gray) did not use a majority-rules criterion but instead leaned more strongly toward voting favorably. This person, let us called her Alberta, would change to a no only if her immediate neighbors were unanimously opposed; if even one favored yes on a given round of discussion, Alberta would stick with her yes. This single biased individual has a big effect in this case, so that the neighborhood eventually stabilizes into two camps in Ending 2—the south camp favorable, and the north unfavorable. In the lower right picture (Ending 3), we see what would happen if there is only one more favorably biased person, let us call her Agnes (also marked in gray) in the north part of the neighborhood. Just these two people completely reversed the emergent neighborhood consensus.

Note that although the outcome was unstable at the beginning, it is very hard to change once the system has stabilized. If Agnes and Alberta were out of town when the first few rounds of discussion happened, the results would have stabilized as in Ending 1, and when Agnes and Alberta returned, they would have both gone along with what was now a unanimous neighborhood consensus against the idea. But after it stabilized favorably, as in Ending 3, unanimous support for the idea would persist even if they both later moved out of the neighborhood.

Where Do the Decision Biases Come From?

In the simulation I showed you in Figure 12.1, the individual neighbors mostly used a simple decision rule: Match the majority of your neighbors. Biological anthropologists Rob Boyd, Pete Richerson, and Joe Henrich have gathered plenty of evidence that human beings are chock-full of conformity mechanisms, which usually serve us quite well (for example, you can randomly try out different leaves and roots for lunch and perhaps die experimenting, or you can eat the ones your neighbors are eating and survive). Psychologists Tanya Chartrand and John Bargh have found that our inclination to imitate other people is automatic and usually unconscious (if you are talking to someone who tends to shake her foot and scratch her eyebrow, you will start shaking your foot and scratching your eyebrow too, without even being aware of it). But as in the case of Agnes and Alberta, some people have higher or lower thresholds for letting others influence them.

In my example, I presumed that people did not have strong preexisting opinions about the outcome. For many of the important decisions that we make in our lives, though, we have strong built-in biases. What determines whether you are favorably disposed to go along with the group or to take an independent stand, to take a dangerous risk or to choose a safe option, to say yes or to say no to an offer to have sex, to fight or to run away in a conflict? To answer those questions, it is critically important to join together the insights of evolutionary psychology with those of dynamical systems theory. As I discussed in earlier chapters, a person's decision biases are predictably different, depending on whether that person is a man or a woman, which stage of life history that person has reached, what his or her chronic mating strategy is, and which of his or her motivational subselves is currently in the driver's seat.

For example, we saw that men whose mating subself is activated are likely to go *against* group opinion, but that women in a mating frame of mind are more likely to go along *with* the group. On the other hand, both men and women conform to group consensus when they are feeling threatened and their night watchman subselves are in charge. Furthermore, our preferred behavioral choices are dynamically linked to the current social situation. If there are a lot of available women as opposed to available men on the horizon, both sexes will adjust their mating strategy, for example. And on a moment-to-moment basis, there is a dynamic connection between the current situation and a person's reaction, so that the brain may switch motivational drivers if what looked initially like a mating opportunity suddenly starts to look a fight.

There is a profoundly important point here. If you want to understand how social complexity arises among humans, you cannot just assume we are like interchangeable ants and that the outcome will depend on random general processes that apply equally to any dynamical system. You have to realize that the outcomes will be powerfully influenced by the particular decision biases that mark our species and that those decision biases are very different depending on which social domain we are currently considering. Finally, individual differences can have a tremendous impact on how it all comes down in the end. I am not merely saying that "things are complex" out there. I am saying that, ironically, we can reduce the complexity by avoiding the tendency to oversimplify. When we combine the insights of dynamical systems theory with those of evolutionary psychology, we get a much richer and more complete understanding of the pattern of social life likely to emerge in the particular kinds of situations we humans create for ourselves. In the next section, I consider how this combination of evolutionary and dynamic ideas yields some important insights, exploring how the different evolved biases inside our heads influence the very shape and structure of our social networks.

Emergent Social Geometries

Although our social networks sometimes look like the flat square neighborhood in Figure 12.1, that is not always the case. In fact, there are different social geometries associated with the different fundamental motivations, as depicted in Figure 12.2. When it comes to status, for example, the geometry is inherently pyramidal, because there are fewer positions available as one goes higher up in the hierarchy. As I have already noted, status has double-barreled payoffs for men, affecting not only their direct access to goodies but also their chances of attracting a desirable mate. For this reason, men are more drawn into these sorts of competitive pyramidal arenas.

The geometry for friendship networks, on the other hand, is flat and relatively permeable: If you are a friend of my brother, then you (and your brother) are likely to be a friend of mine. Females are, in general, more cooperative, and both sexes tend to prefer woman over men as friends. And although it is nice to have friends, truly close friendship networks are inherently limited by our finite budgets of time and energy. There are also coordination problems when the group gets too large: It is tough to throw a picnic if you have to worry about what two hundred people want to eat and drink (part of the reason big weddings are usually a lot more agony than ecstasy).

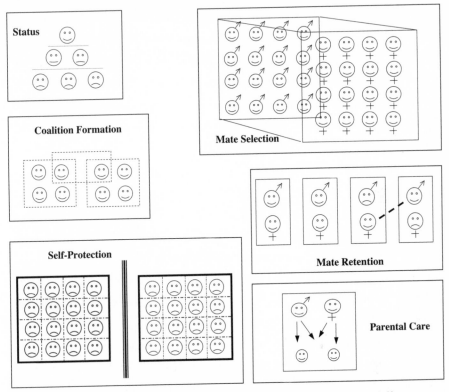

FIGURE 12.2 Emergent social geometries. Our social networks take very different forms depending on whom we are around and which of our subselves is currently in the driver's seat.

When we are worried about self-protection, though, the bigger the gang the better. During my period of hanging around with the hoodlums in the park, there was an actual "gang war" scheduled one night. One of the charming Italian kids from the Forty-sixth Street Boys had made a pass at the wrong guy's sister. The brother in question was a member of an actual Garrisons-style gang, complete with sweaters and jackets emblazoned "Chancellors." Although gang wars had sounded cool when I was eight years old, I had seen that teenage boys could actually draw blood during conflicts and would have sooner been at home reading a book than setting myself up for a trip to the emergency room. But staying home on that particular night would have brought no end of ridicule and harassment. So there I was, trying not to let my hands shake too much and trying in vain to suppress thoughts about what it would feel like to have a chain swung at my head or a knife stuck into my gut. Fortunately, though, every kid in the neighborhood and even some from nearby blocks had been recruited, so I could hope to stay on the periphery when the rumbling started.

As the gang of Chancellors rounded the corner, they were whooping and screaming, working themselves up into a war frenzy. There were a lot of them; we counted about fifty. But our group numbered closer to a hundred. The marauding Chancellors suddenly stopped dead in their tracks, and they must have made a quick visual calculation of the relative size of the two warring parties. After murmuring to one another for a minute or so, they turned around and ran in the other direction. According to Bert Hölldoebler and E. O. Wilson, ants, despite their inability to do arithmetic, have an emergent group-level ability to calculate relative group size, and like these teenage warriors, they are inclined to retreat from an intercolony battle when they discover that they are outnumbered.

Mate selection and mate retention have still different geometries. We select mates from a relatively large population, and are happiest when there are lots of alternatives from which to choose. When Norm Li, Jon Butner, and I surveyed men and women about their mating preferences, we found that the men were slightly more inclined toward promiscuity, whereas the women were strongly inclined toward monogamy. When we plugged men's and women's preferences into a series of dynamic simulations, we found that, over time, the women's preferences tended to win out, so that most simulated neighborhoods ended up predominantly monogamous, with self-perpetuating pockets of promiscuity at the edges. But we also found that the picture changed radically with only small shifts in the women's willingness to forgo monogamy (as you might find in a large city where there are more women than men to go around). Under those circumstances, neighborhoods were more likely to stabilize into a pattern with larger pockets of promiscuity (as could be found in cities like Los Angeles and New York during the 1970s).

Relationship maintenance, on the other hand, is usually a two-person game. Even in polygamous societies, most people end up in couples and not, like Bhupinder Singh, the immensely wealthy seventh maharaja of Patiala, with 350 partners.

If you have as few as two romantic partners, both of them are likely to regard that total as one too many, and each will probably try to drive the other one away. Even in societies with legal polygyny, wives fight with one another over access to the harem leader and over resources for their children. The reason that humans, unlike most other mammals, are so jealous is that both parents contribute resources to the offspring. The woman does not want the man's resources spread too thinly, and the man does not want to contribute resources to another man's child.

Finally, the geometry for kin care is different from the geometry for friendships and mateships in a couple of critical ways. It is not as flat as friendship networks, where resources are shared equitably, but is instead top-down, with resources more easily flowing from parents to children than the reverse. And it is the most stable of all the geometries; romantic partners may leave if things get unfair, but parents are likely to continue providing resources for their children, with many fewer contingencies attached, and parents and children are more motivated to keep up contact even if they are geographically separated.

So the message here is this: Individual human beings have different kinds of decision biases that we carry into each of the different domains of social life. Those biases are not only important in influencing what happens between pairs of people, but they drastically influence the larger web of social life, shaping different social geometries

for our different relationships, and doing so in patterned and functional ways.

It's Emergence and Self-Organization from the Bottom to the Top

Right now, inside your brain, different neural mechanisms are working in parallel to analyze different kinds of sensory inputs about your
body's temperature, your blood sugar levels, about light levels and
noises coming in from the outside, and so on. Our neural systems are
designed to organize themselves so that the most important sensory
inputs get passed up the line for further processing and the others get
ignored. Different kinds of inputs get fed to different subselves, and
the different subselves organize themselves so that only the most critically important one gets to control our consciousness at any given
moment, and thus to make decisions about which specific behavioral
strategy you should pursue next.

When you are in a group of people, each of your behavioral decisions is simultaneously a causal influence on and a calibrated response
to the behaviors of the other people around you. Over time, those mutual interactions organize themselves so that the group moves in a
particular direction.

As in the simulated neighborhoods, most of us interact only lo-
cally, with the closest subset of potential neighbors (even when we are
hooked in to the Internet, we pay attention to a select set of inputs,
such as messages from our friends or from our preferred sources of
news, and ignore most of the others). But because neighbors influence other neighbors, the web of interconnections spreads outward, so
that our decisions and those of the small groups to which we belong
are an integral part of those big anonymous forces like "popular opinion," "modern society," and the "world economy."

Looked at in this light, we can see why human nature is critically
linked to culture, religion, and economics. This is not to say that one

individual's decision biases determine these large-scale phenomena in any unidirectional way, any more than it is to say that "society" determines our individual decisions. Instead, it is to say that broad-scale cultural, historical, and economic patterns emerge from the decisions of individuals.

Most fascinating about this worldview is that it implies there is no Big Brother, no central decision-maker running the show. The emergent aggregate is more powerful and immensely more complex than any single individual. The military-industrial complex, the world economy, public opinion, and modern society is us. The reason it does not seem like it down here is that human society is in some ways like a giant ant colony: a product of many little brains making many little decisions in response to narrow local inputs.

Conclusion

LOOKING UP AT THE STARS

We've come a long way since the 1970s, Baby. The suggestive undercurrents of disco have evolved into the pornographic rhythms of hip-hop, with Jay-Z, T. I., and Eminem jamming to the offspring of the generation that used to shake, shake, shake their booties to KC and the Sunshine Band. Despite the persistently sinful ways of mankind, the Jehovah's Witnesses' prediction that the world would end in 1975 did not come true. It is still a scary place out there, the daily news is still filled with ominous forebodings about economic and political doom, and anxieties about al-Qaida and Iraq have replaced the worries about the Weathermen and Vietnam. But *Homo sapiens* is still around. Fewer and fewer hominids still read the daily newspaper; the bad news is now delivered electronically, via devices called iPhones and BlackBerries—powerful minicomputers smaller than the cigarette packs that used to sit inside the Forty-sixth Street Boys' T-shirt pockets back in 1961.

In my little corner of the world, there are no bombs falling, and the background music is a pleasant Brahms waltz. Contrary to the predictions of the Jesuit priest who counseled me when I was being expelled from Regis High School, I did not end up in a mental institution. Contrary to family tradition, I did not follow my biological father to New York's infamous prison in Ossining. And contrary to my own neurotic fears, I did not get murdered before reaching age fifty.

Instead, I have for several decades lived the comfortable life of a university professor: bicycling to my office down sunny streets with palm trees and swimming pools in the background. I still occasionally wonder when someone in authority will realize there has been a big mistake, but so far the university has kept on giving me paychecks—for doing some of the same things that got me in trouble when I was a high school student: talking too much; making wisecracks in the classroom; sneaking off to read a book whenever I have a lot of overdue assignments; and daydreaming about sex, murder, and the meaning of life. And although I keep discovering more and more things I don't know, I have a better view of the big picture than the young guy who stole away to the bookstore when he should have been studying for his comprehensive exams.

Well then, you may be wondering at this point, what is the meaning of life?

There are, as it turns out, two very different ways to interpret that question.

As I noted in the book's Introduction, the question of the "meaning of life" is sometimes taken to mean "How does it all fit together?" How do the diverse perceptions, thoughts, and feelings buzzing around inside our overgrown brains combine with our prehensile thumbs, upright postures, and chattering larynxes to produce one coordinated organism, and how does our walking, talking species fit into the natural world alongside spider monkeys, timber wolves, and red-faced warblers? I think we now have a pretty good answer to that question, and it has come from integrating the insights of evolutionary psychology with those of cognitive science and dynamical systems theory.

But sometimes when people ask the question about the meaning of life, they are really asking, "How can I live a more meaningful life?" As I observed earlier, academic intellectuals who think Big Thoughts about evolutionary biology, neural networks, and multidirectional causality usually leave this "how-to" question to the pop psych gurus. But I think there are some important connections between the two versions of the "meaning of life" question.

The Meaning of Life I

Let us start with the easy one: How does it all fit together? How do your momentary thoughts about sex, homicide, modern art, and Wall Street link up with one another, and how are those passing thoughts connected to your long-term life choices about career, marriage, family, and religion? Zooming out, how do what is going on inside your head and the choices you make connect with the thoughts and behaviors of your family members, your coworkers, your neighbors, and the billions of strangers in Boise, Brussels, and Beijing? Taking the aerial view, how do today's historical, cultural, and economic events link up with what our ancestors were doing in the Altamira Cave and the Olduvai Gorge? And at the widest camera angle, how does the behavior of us book-reading hominids link up with the behavior of howling baboons, preening peacocks, and mindless colonies of army ants?

Those who specialize too much, as scientists did for much of the twentieth century, might be inclined to think there is not much to be gained from asking such broad questions. But as psychologists, biologists, anthropologists, economists, and other behavioral scientists have increasingly ventured out of their intellectual ghettos, we have begun to discover some fascinating patterns at the interdisciplinary boundaries. As I have discussed in earlier chapters, there have been three big interdisciplinary movements in the last few decades: cognitive science, evolutionary psychology, and dynamical systems theory. Let me try to distill a few take-home lessons from each of these three sets of very big ideas.

Inspired by the discovery that machines could perform some of the functions of the human brain, cognitive scientists rejected the strictures of radical behaviorism and tried to shine a light inside the black box of the mind itself. One of the most important things they discovered is that our brains process information in incredibly selective ways: We only pay attention to a small fraction of what is going on out there, we only ruminate consciously about an even smaller fraction, and we upload a still smaller portion into our long-term memory stores. Which information we select and which we throw out can have big downstream consequences. As I discussed in Chapter 2, to give a simple example, men looking at a crowd stare selectively at the beautiful women, which leads them to later misjudge how prevalent those movie starlets are in the real world, which in turn leads them to feel less committed to their current romantic partners.

As I discussed in Chapter 8, the view of the mind as a computer led cognitive scientists to assume that one form of information processing is interchangeable with any other and to emphasize *process* (how our minds crunch information) over *content* (what information our minds select for crunching in the first place). One of the key insights of evolutionary psychology was that content makes a giant difference; our brains come with some built-in crunching priorities, and they crunch information about poisonous foods, about dangerous predators, and about potential mates in fundamentally different ways. Evolutionary history makes a big difference to which information gets processed and how, as in the differences between bats processing sound waves bouncing off trees to decide which direction to fly, rats processing fragrant chemical compounds to determine whether or not to eat a piece of cheese, and humans processing light waves reflected from another person's facial features to decide whether or not that other person is a friend, an enemy, or a potential mate. Most of the research I talked about in this book reflects a fusion of ideas from cognitive science and evolutionary psychology, as does the view of the mind as being composed of multiple motivational subselves, each designed to deal with different problems.

Besides expanding the focus of psychologists to include content as well as process, the evolutionary perspective has mined some of the powerful general principles from evolutionary biology. The most powerful of those principles—inclusive fitness, differential parental investment, and reciprocal altruism—highlight the connections among such diverse behaviors as aggression, altruism, mate selection, artistic display, conspicuous consumption, and even religious beliefs. They also illuminate the intimate links between our behavior and the behavior of other animals, from nasty gutter rats to elegant peacocks. These connections are still mostly unexplored, and the full implications of these and other broad biological principles are still revealing themselves. In Chapter 7, for example, I talked about life history theory, which has profound implications for understanding how our different motivational subselves make different trade-offs as we develop, shifting priorities in ways designed to maximize successful reproduction. When I was a thirteen-year-old punk hanging around with the Forty-sixth Street Boys, my first priority was status and acceptance; when I was a long-haired college student, my first priority was finding women who would sleep with me; and now as someone approaching eligibility for Social Security, my first priority is helping my children.

These broad ideas have profound consequences not only for psychology but for every other behavioral discipline, including economics, marketing, management, political science, and the law. When the next generation of researchers in these fields overcomes the remaining resistance to thinking about humans as biological organisms, there will be a tidal wave of new discoveries.

Like cognitive science, evolutionary psychology is not a set of inviolable doctrines but a scientific work in progress, as illustrated by the fact that the two broad perspectives are still expanding to incorporate ideas from each other. When psychologists first began testing evolutionary hypotheses about human behavior, we were accused of placing too heavy an emphasis on sex differences and sexual behavior. The accusation was true, because the powerful theories of sexual selection and differential parental investment generated a host of testable predictions about sex differences in mating behavior (which had been surprisingly ignored by psychologists). But science is a self-correcting process, so the accusations spurred a host of studies examining the many ways in which men and women are psychologically similar. Another concern about evolutionary psychology was that it focused too much on the dark and selfish side of human nature. That concern spurred evolutionary psychologists to examine the circumstances under which people act in cooperative, group-oriented, and even heroically altruistic ways.

All this research has led to another important insight: Selfish genes do not necessarily produce selfish people. It is true that our minds are equipped with a host of simple selfish mechanisms that incline us to make decisions promoting our individual reproductive success. But in our ultrasocial species, the goal of reproductive success is often achieved by being nice to others. I will talk more about that shortly.

Although our individual heads are full of simple selfish mechanisms tuned to narrow local inputs, we somehow manage to come together to form smoothly functioning groups, organized societies, and international economic markets. How does this transformation from separate individuals to interconnected groups happen? Enter dynamical systems theory. As I discussed in Chapter 12, research on complex multicomponent systems, such as ant colonies and human neighborhoods, has led to some fascinating insights. One is the discovery of self-organization, or the spontaneous emergence of order out of initial disarray. Another is the discovery that tremendous complexity can emerge from a few simple parameters. Natural selection is itself a stunning example of both these principles, with initially random vari-

ations providing the basis for the emergence of complex living organisms exquisitely adapted to their natural habitats, like night-flying
mammals who use sound to see in the dark, long-beaked birds that
can hover like helicopters as they drink nectar from deep-stemmed
flowers, and naked apes that can communicate complex ideas about
bats, hummingbirds, and the meaning of life.

When we join together the ideas of dynamical systems theory with
those of evolutionary psychology, we begin to understand why different social geometries and different forms of social stability emerge
from the different biases that guide interactions between parents and
children, between lovers, between coworkers, between friends, and between tribes of faceless strangers.

There is a lot more that could be said here, but it will suffice to
say that the field of psychology, which was in a shaky theoretical
state in the years leading up to 1975, has now been revolutionized.
And although none of my fellow troublemakers has yet been elected
president of the American Psychological Association, à la Nelson
Mandela, we have at least been released from our intellectual prison
cells and have been enjoying freedom of the press. Sure, there are still
those who want to cling to one or another version of the blank slate
view, who fall prey to the naturalistic fallacy, or who prefer the more
empirically manageable strictures of an analytic and linear approach
to science. But for the most part, the revolution is over, and we are
in the process of rebuilding a more unified and balanced republic of
behavioral science.

The Meaning of Life II

After I had written several chapters for this book, I learned that I was
going to be featured on an episode of *The Oprah Winfrey Show*. I had
been in a popular documentary with the alluring title *Science of Sex*. I'm
guessing that Oprah or one of her assistants had seen that documentary

and figured the topic would interest her audience, many of whom tune in to hear psychologists, pop stars, and regular people talk about how to overcome everyday problems, have satisfying relationships, and live a more fulfilling life. I started pondering whether the ideas in this book had anything to say about the version of the "meaning of life" question that most of her viewers probably found more relevant: How do you live a more meaningful life?

As I pondered the Oprah question, I was making lunch for my younger son to take to his preschool and calculating how I might squeeze in some time with my older son and his two kids before getting to work that day. I had, as usual, a backlog of work to do; I had agreed to edit a book dedicated to my mentor Bob Cialdini, I had a half dozen research papers to write, and more to revise. Besides all that, there were overdue reviews to do for journal editors, letters of recommendation to write for my graduate students, and decisions to make about upcoming grants—and that was just the job-related list. At home, my tax return was overdue, the roof had a serious leak, and small mammals moving into my attic had to be evicted forthwith. With all I had going on, I should have been miserably anxious, but surprisingly, I was feeling rather calm and focused—enjoying the process of spreading cashew butter on a piece of wheat bread and taking care to trim the crust to my younger son's precise and demanding specifications. And then it struck me: Whatever else has been going on in my life, my children's needs have always trumped all other demands. As I began to think about it, I realized that all this research on people's simple and selfish biases really did have something to say about how to live a more meaningful life.

Unlike many of the other things I have done to seek pleasure, the time I've spent with either of my sons has never given me the slightest hangover of regret. Although it has not always made me euphorically happy to respond to their needs, it is the one thing in my life that reliably makes me feel truly fulfilled. I should not have been surprised, given what I know about evolution and behavior. Human beings are ultimately designed not to seek ecstatic happiness from dawn to dusk but to be linked into a supportive web with other human beings. Indeed, two of the bedrock principles of evolutionary biology are kin selection and reciprocal altruism. The first explains why we are driven to take care of our family members; the second clarifies why we often go to such great lengths to do favors for our friends and coworkers. And although the other central principles, sexual selection and differential parental investment, are typically considered as they relate to showing off and having sex, they are, for human males and females alike, intimately connected to the process of qualifying to be a parent.

I am not suggesting that we all ought to go forth and multiply, ignoring the problem of overpopulation, or that you rush out to make five hundred new Facebook "friends." What I am suggesting instead is that you let yourself enjoy the natural pleasures of taking care of the intimate associates you already have. You can regard time spent with family and friends as a distraction from the central task of life, or you can slow down and let your brain's social mechanisms savor the experiences.

Ask Not What You Can Do for Yourself

This is not a self-help book, so I won't try to work up a list of ten rules for personal fulfillment, rules like "Stick by your family," "Love the one you're with," and "Be loyal to your team members." But I will tell you about the worst piece of advice I heard in my life. I heard it when I was moving toward a divorce from my first wife: "You've got to do what's right for you." I heard it again and again from different people, and even at the time I wondered how doing what was right for me could also be right for my young son. Like many others before and since, I learned the hard way that the mantra should have been "You've got to do what's right for those you love."

While I was avoiding the serious work of finishing up this book, I snuck off to the bookstore and bought a book called *The How of Happiness* by my friend Sonja Lyubomirsky. Unlike a lot of self-help books, Sonja's book offers advice based on rigorous research by a new generation of researchers in a field called "positive psychology." And guess what? Their research suggests that, although family members and friends can often be demanding and annoying on a moment-to-moment basis, people who spend time doing nice things for their loved ones are ultimately less depressed and more fulfilled in their own personal lives.

My favorite positive psychology study is one published in *Science* by Liz Dunn and her colleagues at UBC. They found that people who spent their salary bonuses on other people were happier than those who spent it on themselves. And they did an experiment in which they gave students five dollars or twenty dollars and instructed them to either spend the money on themselves or on someone else. Like rational economists, other students guessed that it would make people happiest to get the larger amount and to spend it on themselves. But that is not what happened. Instead, the students were happiest when they bought someone else a gift, regardless of the amount. These findings are part of a heartening wave of new research suggesting that human beings are chock-full of mechanisms designed to make us feel good when we cement our bonds with those around us.

This Is Dedicated to the Ones I Love

It is customary to put the dedication at the beginning of a book, and it is often a formality, with no real connection to the book that follows. But given what I've just been talking about, it makes the most sense to put the dedication right here: to my wife Carol Luce, and my two sons, Liam, age six, and David, age thirty-two. My wife, besides keeping me nurtured on foods prepared to the precise standards of *Cook's Illustrated*, joined with her wonderfully supportive mother,

Jean Luce, to sponsor me for the long periods I spent holed up in my summer house, neglecting parental duties while writing this book.

It is also appropriate to mention here my longtime collaborators and friends—Rich Keefe, Bob Cialdini, Mark Schaller, Melanie Trost, Ed Sadalla, Sara Gutierres, and Steve Neuberg. And then there is a whole new generation of former students and current collaborators who, as you have seen in the research I described in this book, are now carrying me on their backs, including Jill Sundie, Norm Li, Vlad Griskevicius, Jon Maner, Vaughn Becker, Josh Ackerman, and Jessica Li.

John Alcock, one of my favorite science writers, not only allowed me to sit in on his scientific writing class, but then was generous enough to go through an entire draft of this book and provide me with immensely detailed and helpful advice. David Lundberg Kenrick read through the book with the talented eye of a screenwriter. My wife Carol Luce not only supported me during the long periods I spent writing but also then gave it a reading with her unique twin expertise as a statistical analyst and wife.

And then there is my first wife, Elaine Lundberg, who suffered through my years as a testosterone-crazed young man and who now collaborates with me to help take care of our two grandchildren. In different ways, all these people showed me firsthand how the human mind is wired up to connect us to other human beings and helped me understand the two meanings of the meaning of life.

As the credits roll, we fade to the sun going down over a palm-dotted landscape. A gray-haired professorial type, distinguished in appearance except for his big nose and big feet, is riding an upright city-style bicycle. He is flanked by his handsome adult son on a mountain bike and a cute little blonde-headed boy on a smaller bicycle with training wheels. There is music playing faintly in the background. And in the end, the Beatles are singing, "The love you take is equal to the love you make."

Notes

Introduction: You, Me, Charles Darwin, and Dr. Seuss

viii   *You, me, Jennifer Lopez:* For an overview of ideas from complexity theory, see Nowak & Vallacher, 1998; for an easy introduction to cognitive science, see Gardner, 1985. For some discussion of the bridges between these disciplines and evolutionary psychology, see Kenrick, 2001, and Kenrick, Li, & Butner, 2003.

x   *Simple selfish rules:* For an overview of the simple selfish rules, see Neuberg, Kenrick, & Schaller, 2010, and Schaller, Park, & Kenrick, 2007. For illustrative demonstrations of application to creativity and the like, see Griskevicius, Cialdini, & Kenrick, 2006, and Weeden, Cohen, & Kenrick, 2008. For research on the nuts-and-bolts aspects of courtship and sex, see N. P. Li & Kenrick, 2006; Kenrick, Sadalla, Groth, & Trost, 1990; or Maner et al., 2003. See Buss, 2007, for an extensive overview of evolution-inspired research on mating and other topics.

x   *Simple rules do not mean simple people:* See Kenrick, Griskevicius, Neuberg, & Schaller, 2010.

x   *Simple does not mean irrational:* For a discussion of the concept of deep rationality, see Kenrick, Griskevicius, et al., 2009, and Kenrick, Sundie, & Kurzban, 2008. For related research on conspicuous consumption, see Sundie et al., in press.

xi   *Selfish rules do not create selfish people:* See Kenrick, Sundie, & Kurzban, 2008, and D. S. Wilson, Van Vugt, & O'Gorman, 2008.

xi   *Simple rules unfold into societal complexity:* See Kenrick, Li, & Butner, 2003.

xi   *Procrastination 101:* For good books by the scientific authors I mention, see Alcock, 2001; Buss, 2007; Pinker, 1994; G. F. Miller, 2000; and Lyubomirsky, 2007. The autobiographical books I mention are Bourdain, 1995; Karr, 2005; and Sapolsky, 2002. If you have not had the pleasure of reading Dr. Seuss for a while, your inner child might get a kick out of *If I Ran the Circus* (1956). Douglas Adams's classic is *The Hitchhiker's Guide to the Galaxy* (1979). Mark Twain's *The Prince and the Pauper* (1882) is a great story for young kids, with a clever social commentary running along for older folks who get to read it out loud.

Chapter 1: Standing in the Gutter

2 *My undoing:* The two books that I discuss as the direct triggers to my conversion to an evolutionary perspective are Lancaster, 1976, and E. O. Wilson, 1975. The research that elicited the reviewer's felt duty to protect the "unwary journal readership" was published a decade after that review as Sadalla, Kenrick, & Vershure, 1987, and I will discuss some of the details in Chapter 9.

5 *The academic tumult surrounding sociobiology:* See the collection of readings by Caplan, 1978. For a historical overview, see Segerstråle, 2000. For a sampling of my opinions on some of the controversy, see Kenrick, 1995, 2006b, in press. See also Alcock, 2001, and Pinker, 2002.

5 *The Importance of Being Earnest:* Oscar Wilde's quotation about the gutter and the stars is from *Lady Windermere's Fan* (1892), act 3. My research on homicidal fantasies was published as Kenrick & Sheets, 1994. For evolutionary analyses of themes in cinema, see David Lundberg Kenrick's *Psychology Today* blog *The Caveman Goes to Hollywood*. For a few examples of evolutionary analyses of broader social issues, see Crawford & Salmon, 2004. Applications have included research on prejudice and intergroup conflict (Ackerman et al., 2006; Cottrell & Neuberg, 2005; Kurzban, Tooby, & Cosmides, 2001; Navarette et al., 2009; Schaller, Park, & Mueller, 2003), on sexual harassment (Haselton & Buss, 2000; Kenrick, Trost, & Sheets, 1996), on homicide (Daly & Wilson, 1988), on political conflict (Kurzban, Dukes, & Weeden, 2010; Sidanius & Pratto, 1999; Weeden, Cohen, & Kenrick, 2008), and on environmental problems (Penn, 2003).

Chapter 2: Why *Playboy* Is Bad for Your Mental Mechanisms

11 *Fleeting Glances and Forgettable Faces:* The "booming, buzzing confusion" quotation comes from James, 1890. The research on attention to attractive women and men is reported in Maner et al., 2003. Women's inability to remember attractive men is reported in Maner et al., 2003, and in Anderson et al., 2010. The concentration game research is discussed in Becker, Kenrick, Guerin, & Maner, 2005. For a more detailed discussion of these issues, see Kenrick, Delton, Robertson, Becker, & Neuberg, 2007.

15 *Contrast Effects:* See Helson, 1947, for the historical introduction of his theory of adaptation-level theory and contrast effects. Our initial research studying contrast and physical attractiveness is reported in Kenrick & Gutierres, 1980. The *Playboy* research is reported in Kenrick, Gutierres, & Goldberg, 1989. The finding that women are more influenced by dominance and men by beauty is reported in Kenrick, Neuberg, Zierk, & Krones, 1994.

18 *Comparing Ourselves to Starlets and Moguls:* The effects of such comparisons on self-judgments are reported in Gutierres, Kenrick, & Partch, 1999. For a discussion of parallel effects on people's mood, see Kenrick, Montello,

Gutierres, & Trost, 1993. Much of this goes on at a nonconscious level, analogous to the various effects of hormones on judgment (e.g., Gangestad, Simpson, Cousins, Garver-Apgar, & Christensen, 2004; Haselton & Gangestad, 2006; Penton-Voak et al., 1999; Little, Jones, & DeBruine, 2008; Thornhill & Gangestad, 1999).

Chapter 3: Homicidal Fantasies

24   *Everyday Murderous Thoughts:* Our research on homicidal fantasies is reported in Kenrick & Sheets, 1994. For a discussion of the similar results from the University of Texas, see Buss, 2005b, or Buss & Duntley, 2006. For an extensive treatment of sex differences in actual homicides around the world, see Daly & Wilson, 1988. For a discussion of women's inhibitions against direct aggression, see Björkqvist, Lagerspetz, & Kaukiainen, 1992.

28   *Aggressing to Impress:* For a discussion of Capone's murder of Anselmi, Scalise, and Giunta and of the norms of violence among mafiosi, see Schoenberg, 1992, and Servadio, 1976. The initial research on "trivial altercations" was reported in Wolfgang, 1958. The quotation from the Dallas detective is taken from Mulvihill, Tumin, & Curtis, 1969, p. 230. For a discussion of the nontrivial significance of trivial altercations, see M. Wilson & Daly, 1985. For a good general discussion of sexual selection and differential parental investment, see Daly & Wilson, 1983.

31   *Experimenting with Status-Linked Violence:* For the links among marriage, parenting, and testosterone, see Gray, Chapman, et al., 2004; Gray, Campbell, Marlowe, Lipson, & Ellison, 2004; Gray, Kahlenberg, Barrett, Lipson, & Ellison, 2002; and McIntyre et al., 2006. For a review of findings on the links among social class, testosterone, and aggression, see Dabbs & Morris, 1990, and Rowe, 1996. Our experimental studies of aggression and status motives are reported in Griskevicius et al., 2009.

34   *Levels of Analysis:* See Alcock, 1998, for a discussion of some of Gould's problems with evolutionary analyses of human behavior. For a discussion of levels of analysis, see Sherman, 1988; Simpson & Gangestad, 2001; and Kenrick, Griskevicius, Neuberg, & Schaller, 2010.

37   *When Women Get Direct:* See Campbell, 1999, for an evolutionarily informed analysis of women's aggression. See Cadbury, 2002, for a discussion of the French women's march on the palace at Versailles. For a discussion of the experimental study of women's aggression and resource scarcity, see Griskevicius et al., 2009. Muller, 2007, discusses violence in female chimps. The information about Lizzie Borden's conflict with her father is from Wikipedia (http://en.wikipedia.org/wiki/Lizzie_Borden).

39   *Why Do Men Fantasize About Killing Strangers?:* Schredl, 2009, describes the research on sex differences in aggression in dreams. The research on people's inclination to think of a man when they think of anger is presented in Becker, Kenrick, Neuberg, Blackwell, & Smith, 2007.

Chapter 4: Outgroup Hatred in the Blink of an Eye

43 *A Failure to Discriminate:* For press reports on Lenell Geter's case, see Applebome, 1983, 1984. The quotation from him is from a personal communication to my colleague Steve Neuberg, who interviewed Geter for coverage in our social psychology text's chapter on prejudice (Kenrick, Neuberg, & Cialdini, 2010). The research on how anger reverses outgroup homogeneity is reported in Ackerman et al., 2006. For related research demonstrating enhanced efficiency in processing black men's faces by frightened white participants, see Becker et al., 2010. For research demonstrating that threat can increase stereotyped cognitive processing, see Shapiro et al., 2009.

45 *Functional Projection:* The research demonstrating people's tendencies to project anger onto black men when frightened and men's tendencies to project sexual interest onto attractive women is discussed in Maner et al., 2005. For a related finding, see Haselton & Buss, 2000. The research demonstrating that darkness increases Canadian students' tendencies to see threats in black and Arab men is discussed in Schaller, Park, & Mueller, 2003. The scale measuring belief in a dangerous world is from Altemeyer, 1988. The research on brain activity in response to strange and familiar black men is reported in Phelps et al., 2000.

49 *When Foreign Equals Disgusting:* Paul Rozin's research on disgust and social relationships is discussed in Rozin, Haidt, & McCauley, 2000. The research on foreignness and prejudice is reported in Faulkner, Schaller, Park, & Duncan, 2004. For related research, see Navarrete & Fessler, 2006, and Park, Faulkner, & Schaller, 2003. *Guns, Germs, and Steel* refers to Diamond, 1999. The research on pregnancy and ethnocentrism is reported in Navarrete, Fessler, & Eng, 2007.

52 *Race and Politics:* For research examining different types of prejudice from an evolutionary perspective, see Cottrell & Neuberg, 2005.

54 *The Fallacy of Biology's "Right Wing":* The history of the political controversy surrounding sociobiology can be found in Segerstråle, 2000. The research examining the actual political attitudes of evolutionary psychologists is reported in Tybur, Miller, & Gangestad, 2008. The quotation about the transparent libertarian attack is from Rose & Rose, 2000, p. 8. See Gould & Lewontin, 1979, for an example of their anti-adaptationist writings. The two books I recommend for a discussion of the relevant evidence are Alcock, 2001, and Pinker, 2002.

57 *If You Want to Fight Serpents:* See Kurzban, Tooby, & Cosmides, 2001, for a discussion of the erasability of race in evolutionary terms. The quotation is from Kingsolver, 1996, pp. 8, 9.

Chapter 5: The Mind as a Coloring Book

61 *Zen and the Art of Motorcycle Maintenance:* By Robert Pirsig (1974), is a brilliant novel that takes place largely in Bozeman, Montana, where Pirsig was an assistant professor.

62 *Middle-Aged Gent Seeking College Cheerleader:* The quotations on culture and age preferences come from Presser, 1975, p. 202, and from Peplau & Gordon, 1985, p. 261. The original article on the "trophy wife syndrome" was Connelly, 1989.

63 *Reexamining the Evidence:* The research my colleagues and I conducted on age preferences was reported in Kenrick & Keefe, 1992; Kenrick, Gabrielidis, Keefe, & Cornelius, 1996; Buunk, Dijkstra, Kenrick, & Warntjes, 2001; and Buunk, Dijkstra, Fetchenhauer, & Kenrick, 2002.

66 *Searching for Dirty Old Men Across Times and Cultures:* The research in this section is reported in Kenrick & Keefe, 1992. The data from earlier centuries in Amsterdam are discussed in Kenrick, Nieuweboer, & Buunk, 2010. Other cross-cultural studies we mention include Harpending, 1992, on African pastoralists and Otta, Queiroz, Campos, daSilva, & Silveira, 1998, on Brazilians.

69 *A Startling Exception That Proves the Case:* The Tiwi are discussed in Hart and Pillig, 1960; the quotation is on page 25, emphasis in the original. I consider them in evolutionary perspective in Kenrick, Nieuweboer, & Buunk, 2010.

73 *Blank Slates, Jukeboxes, and Coloring Books:* For a general critique of the blank-slate viewpoint, see Pinker, 2002. For other research on cross-cultural consistencies in human behavior, see Buss, 1989; Jankowiak & Fisher, 1992; Daly & Wilson, 1988; Schmitt et al., 2003. For a discussion of the jukebox metaphor, see Tooby & Cosmides, 1992. For a discussion of the coloring book alternative, see Kenrick, Nieuweboer, & Buunk, 2010. For other research suggesting flexible but adaptive constraints on human behavior, see Gangestad, Haselton, & Buss, 2006.

Chapter 6: Subselves

78 *Simple Minds:* For general discussions of the concept of modularity, see Sherry & Schacter, 1987; Barrett & Kurzban, 2006; Pinker, 1997; Cosmides & Tooby, 1992; and Kenrick, Sadalla, & Keefe, 1998.

78 *Simple Minds and Domain-General Theories:* For a discussion of the reinforcement-affect model of attraction, see Clore & Byrne, 1974. For a brilliant attempt to apply the parsimonious behaviorist theory to all behavior, see Skinner, 1953. For my own research demonstrating more attraction toward people we meet under pleasant circumstances, see Kenrick & Johnson, 1979. The research showing that attractiveness may feel good or bad is reported in Kenrick, Montello, Gutierres, & Trost, 1993. Economic-exchange theories are described in Thibaut & Kelly, 1959, and Hatfield, Traupmann, Sprecher, Utne, & Hay, 1985.

81 *Multiple Minds:* The classic study of conditioning biases is Garcia & Koelling, 1966. The research showing that conditioning works differently in quail and rats is reported in Wilcoxon, Dragoin, & Kral, 1971. The discussion of these findings in relation to different forms of memory in birds can be found in

Sherry & Schacter, 1987. Bat echolocation was documented in Galambos & Griffin, 1941.

85 *Homosexuality and the Modular Mind:* Our research on homosexual's age preferences was reported in Kenrick, Keefe, Bryan, Barr, & Brown, 1995. Other research demonstrating similar preferences in homosexual and heterosexual men is reported in Bailey, Gaulin, Agyei, & Gladue, 1994, and VanderLaan & Vasey, 2008.

88 *Is Friendship Akin to Kinship?:* Our research on the distinct systems involved in responding to friends and kin is reported in Ackerman, Kenrick, & Schaller, 2007. Other research examining reactions to incest includes Lieberman, Tooby, & Cosmides, 2003, 2007, and Fessler & Navarrete, 2004. The finding that people with kinlike features are "trustworthy, but not lustworthy" is reported in DeBruine, 2005.

91 *So How Many Subselves Live Inside Your Head?:* The quotation is from James, 1890, p. 192. For further discussion of these issues, see Kenrick, Griskevicius, Neuberg, & Schaller, 2010; Kenrick & Shiota, 2008; and Neuberg, Kenrick, & Schaller, 2010. For a discussion of encapsulation of different mental systems, see Barrett & Kurzban, 2006. For the original notion of "subselves" in light of research on cognitive science, see Martindale, 1980.

Chapter 7: Reconstructing Maslow's Pyramid

97 *Maslow's famous pyramid of motives:* Maslow introduced his idea of a motivational hierarchy in a paper in *Psychological Review* (1943). He expanded on those ideas in his book on motivation and personality (Maslow, 1970). You can find the iconic pyramid of needs reprinted in modern textbooks in general psychology and organizational behavior (e.g., Kreitner & Kinicki, 2008; Myers, 2009; Nairne, 2003). The behavioral view to which Maslow was reacting is well represented in N. Miller & Dollard, 1941.

99 *Sex and the Meaning of Life:* Our attempt to rebuild the pyramid on a more solid evolutionary foundation is in Kenrick, Griskevicius, Neuberg, & Schaller, 2010.

104 *The Evolutionary Functions of Fundamental Motives:* There is research suggesting that people and other animals might come to crave foods rich in particular nutrients lacking in their diets (e.g., Hughes & Dewar, 1971; Hughes & Wood-Grush, 1971; Rozin & Kalat, 1971). For research suggesting that pregnant women avoid foods that could damage the developing fetus, see Fessler, Eng, & Navarrete, 2005, and Profet, 1992. For research on the specificity of the fear system, see Öhman and Mineka, 2001, or Öhman, Lundqvist, & Esteves, 2001. The research on sharing among the Aché is discussed in K. Hill & Hurtado, 1989. For a discussion of the different systems involved in dealing with friends, families, and lovers, see Kenrick, 2006a, or Ackerman, Kenrick, & Schaller, 2007. For research linking creativity and mating, see Griskevicius, Cialdini, & Kenrick, 2006, which I will discuss in more detail in Chapter 9.

107 *Life History Theory and the Developmental Hierarchy:* For discussion of life history theory, see Kaplan & Gangestad, 2005; Kenrick & Luce, 2000; or Stearns, Allal, & Mace, 2008.

110 *Evolutionism, Humanism, and Positive Psychology:* See Maslow, 1970, for a discussion of the distinctions between humanistic psychology, behaviorism, and Freudian psychology. For some reactions to our renovation of Maslow, see Ackerman & Bargh, 2010; Kesibir, Graham, & Oishi, 2010; Lyubomirsky & Boehm, 2010; and Peterson & Park, 2010.

112 *Proximate Motives:* For a review of the research suggesting that social rejection is registered with the same physiological mechanisms used to register physical pain, see MacDonald & Leary, 2005. For a discussion of later-developing motives that build upon rather than replace earlier ones, see Krebs & Van Hesteren, 1994.

Chapter 8: How the Mind Warps

115 *I have what social psychologists . . . called a "flashbulb memory":* The original paper on flashbulb memories is R. Brown & Kulik, 1971.

118 *Cognitive Science Meets Evolutionary Psychology:* The traditional models of cognitive psychology are well described in Glass & Holyoak, 1986, or Seamon, 1980. For discussion of the different levels of information processing, see Craik & Lockhart, 1972. The argument that similar mental mechanisms are involved in different forms of information processing, from reading words to evaluating other people, can be found in Glass & Holyoak, 1986, and Markus & Zajonc, 1985. In *A Treatise of Human Nature*, written in 1739, David Hume said, "Reason is, *and ought only to be,* the slave of the passions" (emphasis added).

119 *The Evolved Computer Inside Your Head:* The research on people's ability to recognize anger more quickly on a man's face and happiness more quickly on a woman's face is reported in Becker, Kenrick, Neuberg, Blackwell, & Smith, 2007. Darwin's evolutionary analysis of the communication function of emotional expressions can be found in his 1872 classic *The Expression of Emotions in Man and Animals.* The research on change detection and attractive faces is reported in Duncan et al., 2007. For a general discussion of the links between fundamental motives and cognitive processes, see Kenrick, Neuberg, Griskevicius, Becker, & Schaller, 2010.

123 *Remembrances of Things Not Past:* For a general discussion of counterfactual thinking, see Roese & Olson, 1997. The research on sex differences in counterfactual thinking is reported in Roese et al., 2006.

Chapter 9: Peacocks, Porsches, and Pablo Picasso

Veblen discussed conspicuous consumption in *The Theory of the Leisure Class* (1899).

130 *Peacocks and Sexual Selection:* Darwin originally introduced the concept of sexual selection in *The Origin of Species* (1859). The importance of a connection

between parental investment and sexual selection is developed in Trivers, 1972. The quote from the Punjab wildlife official comes from Tandon, 2004.

132   *Dominance and Sexual Attraction in Humans:* The research described in this section is reported in Sadalla, Kenrick, & Vershure, 1987. See Townsend & Levy, 1990a, 1990b, for the research on clothing status and attractiveness. Cross-cultural research demonstrating links between status and a man's success in attracting mates can be found in Buss, 1989; J. Hill, 1984; and Turke & Betzig, 1985.

134   *Flashing the Cash:* The research described in this section is reported in Sundie et al., in press, and Griskevicius et al., 2007.

136   *Creative Genius:* Pinker's arguments about creative abilities as by-products are in *How the Mind Works* (1997). For arguments about creativity and sexual selection, see G. F. Miller's *The Mating Mind* (2000). Our experiments examining the links between mating motives and displays of creativity are reported in Griskevicius, Cialdini, & Kenrick, 2006. For a historical analysis of women who inspired artists, see Prose, 2002.

138   *He's a Rebel:* The classic line-judging conformity experiments are reported in Asch, 1955. Our experiments on conformity and mating motivation are reported in Griskevicius, Goldstein, Mortensen, Cialdini, & Kenrick, 2006.

142   *What About Female Displays?:* The findings on women's creativity displays are reported in Griskevicius, Cialdini, & Kenrick, 2006. Results demonstrating that romantic motives increase women's displays of benevolence are reported in Griskevicius, Tybur et al., 2007.

143   *Peacocks, Porsches, and the Meaning of Life:* The research showing that flashy displays are associated with unrestricted sexuality is reported in Sundie et al., in press. The findings on income and sexual attractiveness are reported in Kenrick, Sundie, Nicastle, & Stone, 2001.

Chapter 10: Sex and Religion

145   *I am describing a scene from a movie:* The Robert De Niro movie filmed on the corner where I grew up was *A Bronx Tale*, although the corner was actually in Queens.

148   *The Psychology of Belief and Disbelief:* The different evolutionary analyses of religion I discuss can be found in Boyer, 2003; Atran & Norenzayan, 2004; Kirkpatrick, 2005; D. S. Wilson, 2002; Johnson & Bering, 2006; and Sharif, Norenzayan, & Henrich, 2010.

150   *Reproductive Religiosity:* The research described in this section is reported in Weeden, Cohen, & Kenrick, 2008. The research on ovulation and attraction to handsome men is reported in Haselton & Gangestad, 2006; Penton-Voak et al., 1999; Little, Jones, & DeBruine, 2008; and Thornhill & Gangestad, 1999.

154   *How Flexible Is the Link Between Religiosity and Reproduction?:* The research discussed in this section is reported in Y. J. Li, Cohen, Weeden, & Kenrick,

2010. See Masters & Johnson, 1970, for their discussion of the hypothesized link between sexual inhibition and religious training.

Chapter 11: Deep Rationality and Evolutionary Economics

161 *Economic Selfishness, Psychological Irrationality, and Deep Rationality:* For an engaging discussion of research on behavioral economics, see Ariely, 2009. For the distinction between "Econs" and "Humans," see Thaler & Sunstein, 2008. For a discussion of loss aversion, see Kahneman & Tversky, 1991. For a classic paper introducing this concept, see Kahneman & Tversky, 1979.

163 *Econs, Humans, and Morons:* For a discussion of how "simple heuristics make us smart," see Todd & Gigerenzer, 2007, or the edited volume by Gigerenzer & Selten (2001). For our discussion of the notion of deep rationality, see Kenrick et al., 2009.

164 *How to Make the Prisoner's Dilemma Disappear:* Our discussion of the prisoner's dilemma can be found in Kenrick & Sundie, 2007, and it is developed more broadly in Kenrick, Sundie, & Kurzban, 2008. Some of the ideas were presented in Kenrick, Sanabria, Sundie, & Killeen, 2010.

166 *Life as a Bank Account:* For discussion of life history theory, see Kaplan & Gangestad, 2005; Kenrick & Luce, 2000; or Stearns, Allal, & Mace, 2008.

168 *Shifting Priorities:* See Sundie & Kenrick, 2006; Kenrick, Sundie, & Kurzban, 2008; or Kenrick, Griskevicius, Neuberg, and Schaller, 2010, for a discussion of these issues. For the research exploring sex differences in what counts as a luxury versus a necessity in a mate, see N. P. Li & Kenrick, 2006, or N. P. Li, Bailey, Kenrick, & Linsenmeier, 2002.

170 *Loss Aversion by Morons Versus Loss Aversion by Evols:* The statement "There has been so much research on loss aversion . . ." comes from Vohs & Luce, 2010, p. 736. The research on loss aversion was discussed in Y. J. Li & Kenrick, 2010. Dan Ariely's 2009 book *Predictably Irrational* is a good introduction to his research and the modern behavioral economic perspective.

Chapter 12: Bad Crowds, Chaotic Attractors, and Humans as Ants

177 *The Wrong Crowd:* For some examples of Latané's work applying dynamical ideas to social psychology, see Latané, 1996; Latané & L'Herrou, 1996; and Latané & Nowak, 1994. I am less familiar with Van Orden's published work on this, and much of it flies well over my head, but see Van Orden, 2002, or Van Orden, Holden, & Turvey, 2003.

180 *Chaotic Attractors and the Revenge of the Nerds:* There are a number of very accessible general treatments of dynamical models of behavior (e.g., Holland, 1998; Lewin, 1992; Waldrop, 1992). For other applications to social psychology, see Nowak, Vallacher, Tesser, & Borkowski, 2000, and Nowak & Vallacher, 1998. The other books to which I refer in this section are Hölldoebler & Wilson, 2008; Carroll, 2005; and Capra, 1997.

183 *Self-Organization: Order out of Randomness:* Our research on the emergence of social norms from individual decision rules is presented in Kenrick, Li, & Butner, 2003.

186 *Where Do the Decision Biases Come From?:* For discussions of the prevalence of human conformity mechanisms, see Henrich & Boyd, 1998, and Richerson & Boyd, 1998. See Chartrand & Bargh, 1999, for the research on non-verbal mimicry.

188 *Emergent Social Geometries:* For further discussion of emergent social geometries, see Kenrick et al., 2002. See Hölldoebler & Wilson, 1994, for a discussion of ants' intercolonial battles.

192 *It's Emergence and Self-Organization from the Bottom to the Top:* For a more detailed discussion of these issues, see Kenrick et al., 2002.

Conclusion: Looking Up at the Stars

197 *The Meaning of Life I:* Most of the material in this section was discussed in earlier chapters. See also Kenrick, in press, and Kenrick et al., 2002.

201 *The Meaning of Life II:* The book honoring my mentor Bob Cialdini is Kenrick, Goldstein, & Braver, in press.

203 *Ask Not What You Can Do for Yourself:* The book I mention here is Lyubomirsky, 2007. Two other excellent, but very different, books on the psychology of happiness are Gilbert, 2007, and Haidt, 2006. The research on money and happiness is Dunn, Aknin, & Norton, 2008. Another study in this vein suggests that people who help others may live longer (S. L. Brown, Nesse, Vinokur, & Smith, 2003).

204 *This Is Dedicated to the Ones I Love:* Jack Eurich, Noah Goldstein, Mark Schaller, Bob Cialdini, Sonja Lyubomirsky, Rich Keefe, David Funder, Jason Weeden, Adam Cohen, and Vlad Griskevicius also read all or part of the book and fed me not only helpful servings of feedback but also hefty portions of encouragement.

　　　The book might not have gotten off the ground if not for Dan Ariely's kind introduction to his agent Jim Levine, who provided not only skillful representation but also plenty of sage advice. T. J. Kelleher at Basic Books helped me navigate to the destination airport, doing an amazingly careful and thoughtful job editing the whole text.

References

Ackerman, J., Kenrick, D. T., & Schaller, M. (2007). Is friendship akin to kinship? *Evolution & Human Behavior, 28*, 365–374.

Ackerman, J., Shapiro, J. R., Neuberg, S. L., Kenrick, D. T., Becker, D. V., Griskevicius, V., Maner, J. K., & Schaller, M. (2006). They all look the same to me (unless they're angry): From out-group homogeneity to out-group heterogeneity. *Psychological Science, 17*, 836–840.

Ackerman, J. M., & Bargh, J. A. (2010). The purpose-driven life: Commentary on Kenrick et al. (2010). *Perspectives on Psychological Science, 5*, 323–326.

Adams, D. N. (1979). *The hitchhiker's guide to the galaxy*. London, England: Pan Books.

Alcock, J. (1998). Unpunctuated equilibrium in the *Natural History* essays of Stephen Jay Gould. *Evolution & Human Behavior, 19*, 321–336.

Alcock, J. (2001). *The triumph of sociobiology*. New York, NY: Oxford University Press.

Altemeyer, B. (1988). *Enemies of freedom*. San Francisco, CA: Jossey-Bass.

Anderson, U. S., Perea, E. F., Becker, D. V., Ackerman, J. M., Shapiro, J. R., Neuberg, S. L., & Kenrick, D. T. (2010). I only have eyes for you: Ovulation redirects attention (but not memory) to attractive men. *Journal of Experimental Social Psychology, 46*, 804–808.

Applebome, P. (1983, May 31). Racial issues raised in robbery case. *New York Times*, p. A14.

Applebome, P. (1984, March 22). Black is cleared by new arrest in Texas holdup. *New York Times*, p. A16.

Ariely, D. (2009). *Predictably irrational: The hidden forces that shape our decisions*. New York, NY: HarperCollins.

Asch, S. E. (1955). Opinions and social pressure. *Scientific American, 193*, 31–35.

Atran, S., & Norenzayan, A. (2004). Religion's evolutionary landscape: Counterintuition, commitment, compassion, communion. *Behavioral & Brain Sciences, 27*, 713–770.

Bailey, J. M., Gaulin, S., Agyei, Y., & Gladue, B. A. (1994). Effects of gender and sexual orientation on evolutionarily relevant aspects of human mating psychology. *Journal of Personality and Social Psychology, 66*, 1081–1093.

Barrett, H. C., & Kurzban, R. (2006). Modularity in cognition: Framing the debate. *Psychological Review, 113,* 628–647.

Becker, D. V., Anderson, U. S., Neuberg, S. L., Maner, J. K., Shapiro, J. R., Ackerman, J. M., Schaller, M., & Kenrick, D. T. (2010). More memory bang for the attentional buck: Self-protection goals enhance encoding efficiency for potentially threatening males. *Social Psychological & Personality Science, 1,* 182–189.

Becker, D. V., Kenrick, D. T., Guerin, S., & Maner, J. K. (2005). Concentrating on beauty: Sexual selection and sociospatial memory. *Personality & Social Psychology Bulletin, 31,* 1643–1652.

Becker, D. V., Kenrick, D. T., Neuberg, S. L., Blackwell, K. C., & Smith, D. M. (2007). The confounded nature of angry men and happy women. *Journal of Personality and Social Psychology, 92,* 179–190.

Björkvist, K., Lagerspetz, K. M. J., & Kaukiainen, A. (1992). Do girls manipulate and boys fight? Developmental trends in regard to direct and indirect aggression. *Aggressive Behavior, 18,* 117–127.

Bourdain, A. (1995). *Kitchen confidential: Adventures in the culinary underbelly.* New York, NY: Bloomsbury.

Boyer, P. (2003). Religious thought and behaviour as by-products of brain function. *Trends in Cognitive Science, 7,* 119–124.

Brown, R., & Kulik, J. (1971). Flashbulb memories. *Cognition, 5,* 73–99.

Brown, S. L., Nesse, R. M., Vinokur, A. D., & Smith, D. M. (2003). Providing social support may be more beneficial than receiving it: Results from a prospective study of mortality. *Psychological Science, 14,* 320–327.

Buss, D. M. (1989). Sex differences in human mate preferences: Evolutionary hypotheses tested in 37 cultures. *Behavioral & Brain Sciences, 12,* 1–49.

Buss, D. M. (2005a). *Handbook of evolutionary psychology.* New York, NY: Wiley.

Buss, D. M. (2005b). *The murderer next door: Why the mind is designed to kill.* New York, NY: Penguin.

Buss, D. M. (2007). *Evolutionary psychology: The new science of mind* (3rd ed.). Boston, MA: Allyn & Bacon.

Buss, D. M., & Duntley, J. D. (2006). The evolution of aggression. In M. Schaller, J. A. Simpson, & D. T. Kenrick (Eds.), *Evolution and social psychology* (pp. 263–286). New York, NY: Psychology Press.

Buunk, B. P., Dijkstra, P., Fetchenhauer, D., & Kenrick, D. T. (2002). Age and gender differences in mate selection criteria for various involvement levels. *Personal Relationships, 9,* 271–278.

Buunk, B. P., Dijstra, P., Kenrick, D. T., & Warntjes, A. (2001). Age differences in preferences for mates are related to gender, own age, and involvement level. *Evolution & Human Behavior, 22,* 241–250.

Cadbury, D. (2002). *The lost king of France.* New York, NY: St. Martin's Press.

Campbell, A. (1999). Staying alive: Evolution, culture, and women's intrasexual aggression. *Behavioral & Brain Sciences, 22,* 203–252.

Caplan, A. L. (1978). *The sociobiology debate: Readings on ethical and scientific issues.* New York, NY: Harper & Row.

Capra, F. (1997). *Web of life: A new scientific understanding of living systems*. New York, NY: Anchor Books.

Carroll, S. B. (2005). *Endless forms most beautiful: The new science of evo-devo*. New York, NY: Norton.

Chartrand, T. L., & Bargh, J. A. (1999). The chameleon effect: The perception-behavior link and social interaction. *Journal of Personality and Social Psychology, 76*, 893–910.

Clore, G. L., & Byrne, D. (1974). A reinforcement affect model of attraction. In T. L. Huston (Ed.), *Foundations of interpersonal attraction* (pp. 143–170). New York, NY: Academic Press.

Connelly, J. (1989, August 28). The CEO's second wife. *Fortune, 120*(5), 52+.

Cosmides, L., & Tooby, J. (1992). Cognitive adaptations for social exchange. In J. H. Barkow, L. Cosmides, & J. Tooby (Eds.), *The adapted mind* (pp. 163–228). New York, NY: Oxford University Press.

Cottrell, C. A., & Neuberg, S. L. (2005). Different emotional reactions to different groups: A sociofunctional threat-based approach to "prejudice." *Journal of Personality & Social Psychology, 88*, 770–789.

Craik, F. I. M., & Lockhart, R. S. (1972). Levels of processing: A framework for memory research. *Journal of Verbal Learning & Verbal Behavior, 11*, 671–684.

Crawford, C. B., & Salmon, C. (2004). *Evolutionary psychology, public policy, and personal decisions*. Mahwah, NJ: Erlbaum.

Dabbs, J. M., Jr., & Morris, R. (1990). Testosterone, social class, and anti-social behavior in a sample of 4462 men. *Psychological Science, 1*, 209–211.

Daly, M., & Wilson, M. (1983). *Sex, evolution, and behavior* (2nd ed.). Belmont, CA: Wadsworth.

Daly, M., & Wilson, M. (1988). *Homicide*. New York, NY: deGruyter.

Darwin, C. (1859/1998). *The origin of species*. Oxford, England: Oxford University Press.

Darwin, C. (1872). *The expression of emotions in man and animals*. London, England: Murray.

DeBruine L. M. (2005). Trustworthy but not lust-worthy: Context-specific effects of facial resemblance. *Proceedings of the Royal Society of London, B, 272*, 919–922.

Diamond, J. (1999). *Guns, germs, and steel*. New York, NY: Norton.

Doctor Seuss. (1956). *If I ran the circus*. New York, NY: Random House.

Duncan, L. A., Park, J. H., Faulkner, J., Schaller, M., Neuberg, S. L., & Kenrick, D. T. (2007). Adaptive allocation of attention: Effects of sex and sociosexuality on visual attention to attractive opposite-sex faces. *Evolution & Human Behavior, 28*, 359–364.

Dunn, E. W., Aknin, L. B., & Norton, M. I. (2008). Spending money on others promotes happiness. *Science, 319*, 1687–1688.

Faulkner, J., Schaller, M., Park, J. H., & Duncan, L. A. (2004). Evolved disease avoidance mechanisms and contemporary xenophobic attitudes. *Group Processes & Intergroup Relations, 7*, 333–353.

Fessler, D. M. T., Eng, S. J., and Navarrete, C. D. (2005). Elevated disgust sensitivity in the first trimester of pregnancy: Evidence supporting the compensatory
prophylaxis hypothesis. *Evolution & Human Behavior, 26*, 344–351.

Fessler, D. M. T., & Navarrete, C. D. (2004). Third-party attitudes toward sibling incest: Evidence for Westermarck's hypothesis. *Evolution and Human Behavior, 25*,
277–294.

Galambos, R., & Griffin, D. R. (1941). Obstacle avoidance by flying bats: The cries
of bats. *Journal of Experimental Zoology, 89*, 475–490.

Gangestad, S. W., Haselton, M. G., & Buss, D. M. (2006). Evolutionary foundations
of cultural variation: Evoked culture and mate preferences. *Psychological Inquiry,
17*, 75–95.

Gangestad, S. W., Simpson, J. A., Cousins, A. J., Garver-Apgar, C. E., & Christensen, N. P. (2004). Women's preferences for male behavioral displays change
across the menstrual cycle. *Psychological Science, 15*, 203–207.

Garcia, J., & Koelling, R. A. (1966). Relation of cue to consequence in avoidance
learning. *Psychonomic Science, 4*, 123–124.

Gardner, H. (1985). *The mind's new science: A history of the cognitive revolution.* New
York, NY: Basic Books.

Gigerenzer, G., & Selten, R. (2001). *Bounded rationality: The adaptive toolbox.*
Boston, MA: MIT Press.

Gilbert, D. T. (2007). *Stumbling on happiness.* New York, NY: Vintage Books.

Glass, A. L., & Holyoak, K. J. (1986). *Cognition* (2nd ed.). New York, NY: Random
House.

Gould, S. J., & Lewontin, R. (1979). The spandrels of San Marcos and the pan-
glossian paradigm: A critique of the adaptationist programme. *Proceedings of the
Royal Society of London, B, 205*, 581–598.

Gray, P. B., Campbell, B. C., Marlowe, F. W., Lipson, S. F., & Ellison, P. T. (2004).
Social variables predict between-subject but not day-to-day variation in the
testosterone of U.S. men. *Psychoneuronedocrinology, 29*, 1153–1162.

Gray, P. B., Chapman, J. F., Burnham, T. C., McIntyre, M. H., Lipson, S. F., & Ellison, P. T. (2004). Human male pair bonding and testosterone. *Human Nature,
15*, 119–131.

Gray, P. B., Kahlenberg, S. M., Barrett, E. S., Lipson, S. F., & Ellison, P. T. (2002).
Marriage and fatherhood are associated with lower testosterone in males. *Evolution & Human Behavior, 23*, 193–201.

Griskevicius, V., Cialdini, R. B., & Kenrick, D. T. (2006). Peacocks, Picasso, and
parental investment: The effects of romantic motives on creativity. *Journal of Per-
sonality& Social Psychology, 91*, 63–76.

Griskevicius, V., Goldstein, N., Mortensen, C., Cialdini, R. B., & Kenrick, D. T.
(2006). Going along versus going alone: When fundamental motives facilitate
strategic (non)conformity. *Journal of Personality & Social Psychology, 91*, 281–294.

Griskevicius, V., Tybur, J. M., Gangestad, S. W., Perea, E. F., Shapiro, J. R., & Kenrick, D. T. (2009). Aggress to impress: Hostility as an evolved context-dependent
strategy. *Journal of Personality & Social Psychology, 96*, 980–994.

Griskevicius, V., Tybur, J. M., Sundie, J. M., Cialdini, R. B., Miller, G. F., & Kenrick, D. T. (2007). Blatant benevolence and conspicuous consumption: When romantic motives elicit strategic costly signals. *Journal of Personality & Social Psychology, 93,* 85–102.

Gutierres, S. E., Kenrick, D. T., & Partch, J. (1999). Contrast effects in self assessment reflect gender differences in mate selection criteria. *Personality & Social Psychology Bulletin, 25,* 1126–1134.

Haidt, J. (2006). *The happiness hypothesis: Finding modern truth in ancient wisdom.* New York, NY: Basic Books.

Harpending, H. (1992). Age differences between mates in Southern African pastoralists. *Behavioral & Brain Sciences, 15,* 102–103.

Hart, C. W. M., & Pillig, A. R. (1960). *The Tiwi of North Australia.* New York, NY: Holt.

Haselton, M. G., & Buss, D. M. (2000). Error management theory: A new perspective on biases in cross-sex mind reading. *Journal of Personality & Social Psychology, 78,* 81–91.

Haselton, M. G., & Gangestad, S. W. (2006). Conditional expression of women's desires and men's mate guarding across the ovulatory cycle. *Hormones and Behavior, 49,* 509–518.

Hatfield, E., Traupmann, J., Sprecher, S., Utne, M., & Hay, J. (1985). Equity and intimate relationships: Recent research. In W. Ickes (Ed.), *Compatible and incompatible relationships* (pp. 1–27). New York, NY: Springer-Verlag.

Helson, H. (1947). Adaptation-level as frame of reference for prediction of psychophysical data. *American Journal of Psychology, 60,* 1–29.

Henrich, J., & Boyd, R. (1998). The evolution of conformist transmission and the emergence of between-group differences. *Evolution and Human Behavior, 19,* 215–242.

Hill, J. (1984). Prestige and reproductive success in man. *Ethology & Sociobiology, 5,* 77–95.

Hill, K., & Hurtado, M. (1989). Hunter-gatherers of the new world. *American Scientist, 77,* 437–443.

Holland, J. H. (1998). *Emergence: From chaos to order.* Reading, MA: Addison-Wesley.

Hölldoebler, B., & Wilson, E. O. (1994). *Journey to the ants: A story of scientific exploration.* Cambridge, MA: Harvard University Press.

Hölldoebler, B., & Wilson, E. O. (2008). *Superorganism: The beauty, elegances, and strangeness of insect societies.* New York, NY: Norton.

Hughes, B. O., & Dewar, W. A. (1971). A specific appetite for zinc in zinc-depleted domestic fowls. *British Poultry Science, 12,* 255–258.

Hughes, B. O., & Wood-Grush, D. G. M. (1971). Investigations into specific appetites for sodium and thiamine in domestic fowls. *Physiology and Behavior, 6,* 331–339.

James, W. (1890). *The principles of psychology* (Vol. 1). New York, NY: Holt.

Jankowiak, W. R., & Fischer, E. F. (1992). A cross-cultural perspective on romantic love. *Ethnology, 31,* 149–155.

Johnson, D., & Bering, J. (2006). Hand of God, mind of man: Punishment and cognition in the evolution of cooperation. *Evolutionary Psychology, 4*, 219–233.

Kahneman, D., & Tversky, A. (1979). Prospect theory: An analysis of decision under risk. *Econometrica, 47*(2), 263–291.

Kahneman, D., & Tversky, A. (1991). Loss aversion in riskless choice: A reference dependent model. In D. Kahneman & A. Tversky (Eds.), *Choices, values, and frames* (pp. 143–158). Cambridge, England: Cambridge University Press.

Kaplan, H. S., & Gangestad, S. W. (2005). Life history and evolutionary psychology. In D. M. Buss (Ed.), *Handbook of evolutionary psychology* (pp. 68–95). New York, NY: Wiley.

Karr, M. (2005). *The liars' club: A memoir.* New York, NY: Penguin.

Kenrick, D. T. (1995). Evolutionary theory versus the confederacy of dunces. *Psychological Inquiry, 6*, 56–61.

Kenrick, D. T. (2001). Evolution, cognitive science, and dynamical systems: An emerging integrative paradigm. *Current Directions in Psychological Science, 10*, 13–17.

Kenrick, D. T. (2006a). A dynamical evolutionary view of love. In R. J. Sternberg & K. Weis (Eds.), *The new psychology of love* (pp. 15–34). New Haven, CT: Yale University Press.

Kenrick, D. T. (2006b). Evolutionary psychology: Resistance is futile. *Psychological Inquiry, 17*, 102–108.

Kenrick, D. T. (in press). Evolutionary social psychology: From animal instincts to human societies. In A. Kruglanski, P. A. M. VanLange, & T. Higgins, *Handbook of theory in social psychology.* Newbury Park, CA: Sage.

Kenrick, D. T., Delton, A. W., Robertson, T., Becker, D. V. & Neuberg, S. L. (2007). How the mind warps: A social evolutionary perspective on cognitive processing disjunctions. In J. P. Forgas, M. G. Haselton, & W. Von Hippel (Eds.), *Evolution and the social mind: Evolutionary psychology and social cognition* (pp. 49–68). New York, NY: Psychology Press.

Kenrick, D. T., Gabrielidis, C., Keefe, R. C., & Cornelius, J. (1996). Adolescents' age preferences for dating partners: Support for an evolutionary model of life-history strategies. *Child Development, 67*, 1499–1511.

Kenrick, D. T., Goldstein, N., & Braver, S. L. (in press). *Six degrees of social influence: The science and application of Robert Cialdini.* New York, NY: Oxford University Press.

Kenrick, D. T., Griskevicius, V., Neuberg, S. L., & Schaller, M. (2010). Renovating the pyramid of needs: Contemporary extensions built upon ancient foundations. *Perspectives on Psychological Science, 5*, 292–314.

Kenrick, D. T., Griskevicius, V., Sundie, J. M., Li, N. P., Li, Y. J., & Neuberg, S. L. (2009). Deep rationality: The evolutionary economics of decision-making. *Social cognition, 27*, 764–785.

Kenrick, D. T., & Gutierres, S. (1980). Contrast effects and judgments of physical attractiveness: When beauty becomes a social problem. *Journal of Personality & Social Psychology, 38*, 131–140.

Kenrick, D. T., Gutierres, S. E., & Goldberg, L. (1989). Influence of erotica on ratings of strangers and mates. *Journal of Experimental Social Psychology, 25,* 159–167.

Kenrick, D. T., & Johnson, G. A. (1979). Interpersonal attraction in aversive environments: A problem for the classical conditioning paradigm? *Journal of Personality & Social Psychology, 37,* 572–579.

Kenrick, D. T., & Keefe, R. C. (1992). Age preferences in mates reflect sex differences in mating strategies. *Behavioral & Brain Sciences, 15,* 75–91.

Kenrick, D. T., Keefe, R. C., Bryan, A., Barr, A., & Brown, S. (1995). Age preferences and mate choice among homosexuals and heterosexuals: A case for modular psychological mechanisms. *Journal of Personality & Social Psychology, 69,* 1166–1172.

Kenrick, D. T., Li, N. P., & Butner, J. (2003). Dynamical evolutionary psychology: Individual decision-rules and emergent social norms. *Psychological Review, 110,* 3–28.

Kenrick, D. T., & Luce, C. L. (2000). An evolutionary life-history model of gender differences and similarities. In T. Eckes & H. M. Trautner (Eds.), *The developmental social psychology of gender* (pp. 35–64). Hillsdale, NJ: Erlbaum.

Kenrick, D. T., Maner, J. K., Butner, J., Li, N. P., Becker, D. V., & Schaller, M. (2002). Dynamic evolutionary psychology: Mapping the domains of the new interactionist paradigm. *Personality & Social Psychology Review, 6,* 347–356.

Kenrick, D. T., Montello, D. R., Gutierres, S. E., & Trost, M. R. (1993). Effects of physical attractiveness on affect and perceptual judgment: When social comparison overrides social reinforcement. *Personality & Social Psychology Bulletin, 19,* 195–199.

Kenrick, D. T., Neuberg, S. L., & Cialdini, R. B. (2010). *Social psychology: Goals in interaction* (5th ed.). Boston, MA: Allyn & Bacon.

Kenrick, D. T., Neuberg, S. L., Griskevicius, V., Becker, D. V., & Schaller, M. (2010). Goal-driven cognition and functional behavior: The fundamental motives framework. *Current Directions in Psychological Science, 19,* 63–67.

Kenrick, D. T., Neuberg, S. L., Zierk, K., & Krones, J. (1994). Evolution and social cognition: Contrast effects as a function of sex, dominance, and physical attractiveness. *Personality & Social Psychology Bulletin, 20,* 210–217.

Kenrick, D. T., Nieuweboer, S., & Buunk, A. P. (2010). Universal mechanisms and cultural diversity: Replacing the blank slate with a coloring book. In M. Schaller, S. Heine, A. Norenzayan, T. Yamagishi, & T. Kameda (Eds.), *Evolution, culture, and the human mind* (pp. 257–272). Mahwah, NJ: Erlbaum.

Kenrick, D. T., Sadalla, E. K., Groth, G., & Trost, M. R. (1990). Evolution, traits, and the stages of human courtship: Qualifying the parental investment model. *Journal of Personality, 58,* 97–116. Special issue on Biological Approaches to Personality.

Kenrick, D. T., Sadalla, E. K., & Keefe, R. C. (1998). Evolutionary cognitive psychology: The missing heart of modern cognitive science. In C. Crawford & D. L. Krebs (Eds.), *Handbook of evolutionary psychology* (pp. 485–514). Hillsdale, NJ: Erlbaum.

Kenrick, D. T., Sanabria, F., Sundie, J. M., & Killeen, P. R. (2010, January). *Game theory and social domains: How fitness interdependencies transform strategic decisions.* Paper presented at Society for Personality & Social Psychology, Las Vegas, NV.

Kenrick, D. T., & Sheets, V. (1994). Homicidal fantasies. *Ethology & Sociobiology, 14,* 231–246

Kenrick, D. T., & Shiota, M. N. (2008). Approach and avoidance motivation(s): An evolutionary perspective. In A. J. Elliot (Ed.), *Handbook of approach and avoidance motivation* (pp. 273–288). New York, NY: Psychology Press.

Kenrick, D. T., & Sundie, J. M. (2007). Dynamical evolutionary psychology and mathematical modeling: Quantifying the implications of qualitative biases. In S. W. Gangestad & J. A. Simpson (Eds.), *The evolution of mind: Fundamental questions and controversies* (pp. 137–144). New York, NY: Guilford Press.

Kenrick, D. T., Sundie, J. M., & Kurzban, R. (2008). Cooperation and conflict between kith, kin, and strangers: Game theory by domains. In C. Crawford & D. Krebs (Eds.), *Foundations of evolutionary psychology* (pp. 353–370). New York, NY: Erlbaum.

Kenrick, D. T., Sundie, J. M., Nicastle, L. D., & Stone, G. O. (2001). Can one ever be too wealthy or too chaste? Searching for nonlinearities in mate judgment. *Journal of Personality & Social Psychology, 80,* 462–471.

Kenrick, D. T., Trost, M. R., & Sheets, V. L. (1996). The feminist advantages of an evolutionary perspective. In D. M. Buss & N. Malamuth (Eds.), *Sex, power, conflict: Feminist and evolutionary perspectives* (pp. 29–53). New York, NY: Oxford University Press.

Kesebir, S., Graham, J., & Oishi, S. (2010). A theory of human needs should be human-centered, not animal-centered: Commentary on Kenrick et al. (2010). *Perspectives on Psychological Science, 5,* 315–319.

Kingsolver, B. (1996). *High tide in Tucson.* New York, NY: HarperCollins.

Kirkpatrick, L. A. (2005). *Attachment, evolution, and the psychology of religion.* New York, NY: Guilford Press.

Krebs, D. L., & Van Hesteren, F. (1994). The development of altruism: Toward an integrative model. *Developmental Review, 14,* 1–56.

Kreitner, R., & Kinicki, A. (2008). *Organizational behavior* (8th ed.). New York, NY: McGraw-Hill.

Kurzban, R., Dukes, A., & Weeden, J. (2010). Sex, drugs, and moral goals: Reproductive strategies and views about recreational drugs. *Proceedings of the Royal Society, B, 277,* 3501–3508.

Kurzban, R., Tooby, J., & Cosmides, L. (2001). Can race be erased? Coalitional computation and social categorization. *Proceedings of the National Academy of Sciences, 98,* 15387–15392.

Lancaster, J. B. (1976). *Primate behavior and the emergence of human culture.* New York, NY: Holt.

Latané, B. (1996). Dynamic social impact: The creation of culture by communication. *Journal of Communication, 46,* 13–25.

Latané, B., & L'Herrou, T. (1996). Spatial clustering in the conformity game: Dynamic social impact in electronic groups. *Journal of Personality & Social Psychology, 70*, 1218–1230.

Latané, B., & Nowak, A. (1994). Attitudes as catastrophes: From dimensions to categories with increasing involvement. In R. R. Vallacher & A. Nowak (Eds.), *Dynamical systems in social psychology* (pp. 219–249). San Diego, CA: Academic Press.

Lewin, R. (1992). *Complexity: Life at the edge of chaos.* New York, NY: Macmillan.

Li, N. P., Bailey, J. M., Kenrick, D. T., & Linsenmeier, J. A. (2002). The necessities and luxuries of mate preferences: Testing the trade-offs. *Journal of Personality & Social Psychology, 82*, 947–955.

Li, N. P., & Kenrick, D. T. (2006). Sex similarities and differences in preferences for short-term mates: What, whether, and why. *Journal of Personality & Social Psychology, 90*, 468–489.

Li, Y. J., Cohen, A. B., Weeden, J., & Kenrick, D. T. (2010). Mating competitors increase religious beliefs. *Journal of Experimental Social Psychology, 46*, 428–431.

Li, Y. J., & Kenrick, D. T. (2010, January). *Mating motives make men loss averse.* Paper presented at Society for Personality & Social Psychology, Las Vegas, NV.

Lieberman, D., Tooby, J., & Cosmides, L. (2003). Does morality have a biological basis? An empirical test of the factors governing moral sentiments regarding incest. *Proceedings of the Royal Society of London, B, 270*, 819–826.

Lieberman, D., Tooby, J., & Cosmides, L. (2007). The architecture of human kin detection. *Nature, 445*, 727–731.

Little, A. C., Jones, B. C., & DeBruine, L. M. (2008). Preferences for variation in masculinity in real male faces change across the menstrual cycle: Women prefer more masculine faces when they are more fertile. *Personality & Individual Differences, 45*, 478–482.

Lyubomirsky, S. (2007). *The how of happiness: A scientific approach to getting the life you want.* New York, NY: Penguin.

Lyubomirsky, S., & Boehm, J. K. (2010). Human motives, happiness, and the puzzle of parenthood: Commentary on Kenrick et al. (2010). *Perspectives on Psychological Science, 5*, 327–334.

MacDonald, G., & Leary, M. R. (2005). Why does social exclusion hurt? The relationship between social and physical pain. *Psychological Bulletin, 131*, 202–223.

Maner, J. K., Kenrick, D. T., Becker, D. V., Delton, A. W., Hofer, B., Wilbur, C. J., & Neuberg, S. L. (2003). Sexually selective cognition: Beauty captures the mind of the beholder. *Journal of Personality & Social Psychology, 6*, 1107–1120

Maner, J. K., Kenrick, D. T., Becker, D. V., Robertson, T. E., Hofer, B., Neuberg, S. L., Delton, A. W., Butner, J., & Schaller, M. (2005). Functional projection: How fundamental social motives can bias interpersonal perception. *Journal of Personality & Social Psychology, 88*, 63–78.

Markus, H., & Zajonc, R. B. (1985). The cognitive perspective in social psychology. In G. Lindzey & E. Aronson (Eds.), *Handbook of social psychology* (Vol. 1, pp. 137–230). New York, NY: Random House.

Martindale, C. (1980). Subselves. In L. Wheeler (Ed.), *Review of personality and social psychology* (pp. 193–218). Beverly Hills, CA: Sage.

Maslow, A. H. (1943). A theory of human motivation. *Psychological Review, 50,* 370–396.

Maslow, A. H. (1970). *Motivation and personality* (2nd ed.). New York, NY: Harper & Row.

Masters, W. H., & Johnson, V. E. (1970). *Human sexual inadequacy.* Toronto; New York: Bantam Books.

McIntyre, M., Gangestad, S. W., Gray, P. B., Chapman, J. F., Burnham, T. C., O'Rourke, M. T., & Thornhill, R. (2006). Romantic involvement often reduces men's testosterone levels—but not always: The moderating role of extrapair sexual interest. *Journal of Personality & Social Psychology, 91,* 642–651.

Miller, G. F. (2000). *The mating mind: How sexual choice shaped the evolution of human nature.* London, England: Heinemann.

Miller, N., & Dollard, J. (1941). *Social learning and imitation.* New Haven, CT: Yale University Press.

Muller, M. (2007). Chimpanzee violence: Femmes fatales. *Current Biology, 17,* R365–R366.

Mulvihill, D. J., Tumin, M. M., & Curtis, L. A. (1969). *Crimes of violence* (Vol. 2). Washington, DC: Government Printing Office.

Myers, D. G. (2009). *Psychology in everyday life.* New York, NY: Worth.

Nairne, J. S. (2003). *Psychology: The adaptive mind.* Belmont, CA: Wadsworth.

Navarrete, C. D., & Fessler, D. M. T. (2006). Disease avoidance and ethnocentrism: The effects of disease vulnerability and disgust sensitivity on intergroup attitudes. *Evolution & Human Behavior, 27,* 270–282.

Navarrete, C. D., Fessler, D. M. T., & Eng, S. J. (2007). Elevated ethnocentrism in the first trimester of pregnancy. *Evolution & Human Behavior, 28,* 60–65.

Navarrete, C. D., Olsson, A., Ho, A. K., Mendes, W. B., Thomsen, L., & Sidanius, J. (2009). Fear extinction to an out-group face: The role of target gender. *Psychological Science, 20,* 155–158.

Neuberg, S. L, Kenrick, D. T., & Schaller, M. (2010). Evolutionary social psychology. In S. T. Fiske, D. T. Gilbert, & G. Lindzey (Eds.), *Handbook of Social Psychology* (5th ed., Vol. 2, pp. 761–796). New York, NY: Wiley & Sons.

Nowak, A., & Vallacher, R. R. (1998). *Dynamical social psychology.* New York, NY: Guilford Press.

Nowak, A., Vallacher, R. R., Tesser, A., & Borkowski, W. (2000). Society of self: The emergence of collective properties in self-structure. *Psychological Review, 107,* 39–61.

Öhman, A., Lundqvist, D., & Esteves, F. (2001). The face in the crowd revisited: A threat advantage with schematic stimuli. *Journal of Personality & Social Psychology, 80,* 381–396.

Öhman, A., & Mineka, S. (2001). Fears, phobias, and preparedness: Toward an evolved module of fear and fear learning. *Psychological Review, 108*, 483–522.

Otta, E., Queiroz, R. D. S., Campos, L. D. S., daSilva, M. W. D., & Silveira, M. T. (1998). Age differences between spouses in a Brazilian marriage sample. *Evolution & Human Behavior, 20*, 99–103.

Park, J. H., Faulkner, J., & Schaller, M. (2003). Evolved disease-avoidance processes and contemporary anti-social behavior: Prejudicial attitudes and avoidance of people with physical disabilities. *Journal of Nonverbal Behavior, 27*, 65–87.

Penn, D. J. (2003). The evolutionary roots of our environmental problems: Toward a Darwinian ecology. *Quarterly Review of Biology, 78*, 275–301.

Penton-Voak, I. S., Perrett, D. I., Castles, D. L., Kobayashi, T., Burt, D. M., Murray, L. K., et al. (1999). Menstrual cycle alters face preference. *Nature, 399*, 741–742.

Peplau, L. A., & Gordon, S. L. (1985). Women and men in love: Gender differences in heterosexual relationships. In V. E. O'Leary, R. K. Unger, & B. S. Wallston (Eds.), *Women, gender, and social psychology* (pp. 257–291). Hillsdale, NJ: Erlbaum.

Peterson, C., & Park, N. (2010). What happened to self-actualization? Commentary on Kenrick et al. (2010). *Perspectives on Psychological Science, 5*, 320–322.

Phelps, E. A., O'Connor, K. J., Cunningham, W. A., Funayama, E. S., Gatenby, J. C., Gore, J. C., & Banaji, M. R. (2000). Performance on indirect measures of race evaluation predicts amygdala activation. *Journal of Cognitive Neuroscience, 12*, 729–738.

Pinker, S. (1994). *The language instinct.* New York, NY: HarperCollins.

Pinker, S. (1997). *How the mind works.* New York, NY: Norton.

Pinker, S. (2002). *The blank slate.* New York, NY: Viking Penguin.

Pirsig, R. (1974). *Zen and the art of motorcycle maintenance: An inquiry into values.* New York, NY: William Morrow.

Presser, H. B. (1975). Age differences between spouses: Trends, patterns, and social implications. *American Behavioral Scientist, 19*, 190–205.

Profet, M. (1992). Pregnancy sickness as adaptation: A deterrent to maternal ingestion of teratogens. In J. H. Barkow, L. Cosmides, & J. Tooby (Eds.), *The adapted mind: Evolutionary psychology and the generation of culture* (pp. 327–366). New York, NY: Oxford University Press.

Prose, F. (2002). *The lives of the muses: Nine women and the artists they inspired.* New York, NY: HarperCollins.

Richerson, P. J., & Boyd, R. (1998). The evolution of human ultrasociality. In I. Eibl-Eibisfeldt & F. Salter (Eds.), *Ideology, warfare, and indoctrinability* (pp. 71–95). New York, NY: Berghahn Books.

Roese, N., Pennington, G. L., Coleman, J., Janicki, M., Li, N. P., & Kenrick, D. T. (2006). Sex differences in regret: All for love or some for lust? *Personality & Social Psychology Bulletin, 32*, 770–780.

Roese, N. J., & Olson, J. M. (1997). Counterfactual thinking: The intersection of affect and function. *Advances in Experimental Social Psychology, 29*, 1–59.

Rose, H., & Rose, S. (2000). Introduction. In H. Rose and S. Rose (Eds.), *Alas poor Darwin: Arguments against evolutionary psychology* (pp. 1–13). London: Harmony Books.

Rowe, D. C. (1996). An adaptive strategy theory of crime and delinquency. In J. D. Hawkins (Ed.), *Delinquency and crime: Current theories* (pp. 268–314). New York, NY: Cambridge University Press.

Rozin, P., Haidt, J., & McCauley, C. R. (2000). Disgust. In M. Lewis & J. M. Haviland-Jones (Eds.), *Handbook of emotions* (2nd ed., pp. 637–653). New York, NY: Guilford Press.

Rozin, P., & Kalat, J. W. (1971). Specific hungers and poison avoidance as adaptive specializations of learning. *Psychological Review, 78,* 459–486.

Sadalla, E. K., Kenrick, D. T., & Vershure, B. (1987). Dominance and heterosexual attraction. *Journal of Personality & Social Psychology, 52,* 730–738.

Sapolsky, R. M. (2002). *A primate's memoir: A neuroscientist's unconventional life among the baboons.* New York, NY: Scribner.

Schaller, M., Park, J. H., & Kenrick, D. T. (2007). Human evolution and social cognition. In R. I. M. Dunbar & L. Barrett (Eds.), *Oxford handbook of evolutionary psychology* (pp. 491–504). Oxford, England: Oxford University Press.

Schaller, M., Park, J. H., & Mueller, A. (2003). Fear of the dark: Interactive effects of beliefs about danger and ambient darkness on ethnic stereotypes. *Personality & Social Psychology Bulletin, 29,* 637–649.

Schmitt, D. P., & 118 Members of the International Sexuality Description Project. (2003). Universal sex differences in the desire for sexual variety: Tests from 52 nations, 6 continents, and 13 islands. *Journal of Personality & Social Psychology, 85,* 85–104.

Schoenberg, R. J. (1992). *Mr. Capone.* New York, NY: William Morrow.

Schredl, M. (2009). Sex differences in dream aggression. *Behavioral & Brain Sciences, 32,* 287–288.

Seamon, J. G. (1980). *Memory and cognition: An introduction.* New York, NY: Oxford University Press.

Segerstråle, U. (2000). *Defenders of the truth: The battle for science in the sociology debate and beyond.* Oxford, England: Oxford University Press.

Servadio, G. (1976). *Mafioso: A history of the Mafia from its origins to the present day.* New York, NY: Stein & Day.

Shapiro, J. R., Ackerman, J. M., Neuberg, S. L., Maner, J. K., Becker, D. V., & Kenrick, D. T. (2009). Following in the wake of anger: When not discriminating is discriminating. *Personality & Social Psychology Bulletin, 35,* 1356–1367.

Sharif, A. F., Norenzayan, A., & Henrich, J. (2010). The birth of high gods: How the cultural evolution of supernatural policing influenced the emergence of complex, cooperative human societies, paving the way for civilization. In M. Schaller, S. Heine, A. Norenzayan, T. Yamagishi, & T. Kameda (Eds.), *Evolution, culture, and the human mind* (pp. 119–136). Mahwah, NJ: Erlbaum.

Sherman, P. W. (1988). The levels of analysis. *Animal Behavior, 36,* 616–619.

Sherry, D. F., & Schacter, D. L. (1987). The evolution of multiple memory systems. *Psychological Review, 94,* 439–454.

Sidanius, J., & Pratto, F. (1999). *Social dominance: An intergroup theory of social hierarchy and oppression.* New York, NY: Cambridge University Press.

Simpson, J. A., & Gangestad, S. W. (2001). Evolution and relationships: A call for integration. *Personal Relationships, 8*, 341–355.

Skinner, B. F. (1953). *Science and human behavior.* New York, NY: Free Press.

Stearns, S. C., Allal, N., & Mace, R. (2008). Life history theory and human development. In C. Crawford & D. Krebs (Eds.), *Foundations of evolutionary psychology* (pp. 47–70). New York, NY: Erlbaum.

Sundie, J. M., & Kenrick, D. T. (2006). Modular economics: Different bonds = different investments. *Psychological Inquiry, 17*, 56–59.

Sundie, J. M., Kenrick, D. T., Griskevicius, V., Tybur, J., Vohs, K., & Beal, D. J. (in press). Peacocks, Porsches, and Thorsten Veblen: Conspicuous consumption as a sexual signaling system. *Journal of Personality & Social Psychology.*

Tandon, A. (2004, May 2). Peacock in peril. *Tribune of India.* Retrieved from http://www.tribuneindia.com/2004/20040502/spectrum/main1.htm.

Thaler, R. H., & Sunstein, C. R. (2008). *Nudge: Improving decisions about health, wealth, and happiness.* New Haven, CT: Yale University Press.

Thibaut, J., & Kelley, H. H. (1959). *The social psychology of groups.* New York, NY: Wiley.

Thornhill, R., & Gangestad, S. W. (1999). The scent of symmetry: A human sex pheromone that signals fitness? *Evolution & Human Behavior, 20*, 175–201.

Todd, P. M., & Gigerenzer, G. (2007). Mechanisms of ecological rationality: Heuristics and environments that make us smart. In R. I. M. Dunbar & L. Barrett (Eds.), *Oxford handbook of evolutionary psychology* (pp. 197–210). Oxford, England: Oxford University Press.

Tooby, J., & Cosmides, L. (1992). The psychological foundations of culture. In J. H. Barkow, L. Cosmides, & J. Tooby (Eds.), *The adapted mind: Evolutionary psychology and the generation of culture* (pp. 19–136). New York, NY: Oxford University Press.

Townsend, J. M., & Levy, G. D. (1990a). Effects of potential partner's costume and physical attractiveness on sexuality and partner selection: Sex differences in reported preferences of university students. *Journal of Psychology, 124*, 371–376.

Townsend, J. M., & Levy, G. D. (1990b). Effects of potential partner's physical attractiveness and socioeconomic status on sexuality and partner selection. *Archives of Sexual Behavior, 19*, 149–164.

Trivers, Robert L. (1972). Parental investment and sexual selection. In B. G. Campbell (Ed.), *Sexual selection and the descent of man* (pp. 136–179). Chicago, IL: Aldine.

Turke, P. W., & Betzig, L. L. (1985). Those who can do: Wealth, status, and reproductive success on Ifaluk. *Ethology & Sociobiology, 6*, 79–87.

Twain, M. (1882). *The prince and the pauper.* New York, NY: James Osgood.

Tybur, J. M., Miller, G. F., & Gangestad, S. W. (2008). Testing the controversy: An empirical examination of adaptationists' attitudes toward politics and science. *Human nature: An interdisciplinary biosocial perspective, 18*, 313–328.

VanderLaan, D. P., & Vasey, P. L. (2008). Mate retention behavior of men and women in heterosexual and homosexual relationships. *Archives of Sexual Behavior, 37*, 572–585.

Van Orden, G. C. (2002). Nonlinear dynamics and psycholinguistics. *Ecological Psychology, 14*, 1–4.

Van Orden, G. C., Holden, J. G., & Turvey, M. T. (2003). Self-organization of cognitive performance. *Journal of Experimental Psychology-General, 132*, 331–350.

Veblen, T. (1899/1994). *The theory of the leisure class*. New York, NY: Mentor Books.

Vohs, K. D., & Luce, M. F. (2010). Judgment and decision making. In R. F. Baumeister & E. J. Finkel (Eds.), *Advanced social psychology: The state of science* (pp. 733–756). New York, NY: Oxford University Press.

Waldrop, M. M. (1992). *Complexity: The emerging science at the edge of order and chaos*. New York, NY: Simon & Schuster.

Weeden, J., Cohen, A. B., & Kenrick, D. T. (2008). Religious attendance as reproductive support. *Evolution & Human Behavior, 29*, 327–334.

Wilcoxon, H. C., Dragoin, W. B., & Kral, P. A. (1971). Illness-induced aversions in rat and quail: Relative salience of visual and gustatory cues. *Science, 171*, 826–828.

Wilson, D. S. (2002). *Darwin's cathedral: Evolution, religion, and the nature of society*. Chicago, IL: University of Chicago Press.

Wilson, D. S., Van Vugt, M., & O'Gorman, R. (2008). Multilevel selection theory and major evolutionary transitions: Implications for psychological science. *Current Directions in Psychological Science, 17*, 6–9.

Wilson, E. O. (1975). *Sociobiology: The new synthesis*. Cambridge, MA: Harvard University Press.

Wilson, M., & Daly, M. (1985). Competitiveness, risk taking, and violence: The young male syndrome. *Ethology & Sociobiology, 6*, 59–73.

Wolfgang, M. E. (1958). *Patterns in criminal homicide*. Philadelphia: University of Pennsylvania Press.

Index